IET POWER AND ENERGY SERIES 44

Series Editors: Professor A.T. Johns
Professor D.F. Warne

Economic Evaluation of Projects in the Electricity Supply Industry

Other volumes in this series:

Economic Evaluation of Projects in the Electricity Supply Industry

Hisham Khatib

The Institution of Engineering and Technology

Published by The Institution of Engineering and Technology, London, United Kingdom

First published 1996 (0 85296 908 2)
Revised edition 2003 (0 86341 304 8)

The Institution of Engineering and Technology
Michael Faraday House
Six Hills Way, Stevenage
Herts, SG1 2AY, United Kingdom

www.theiet.org

British Library Cataloguing in Publication Data
Khatib, Hisham, 1936 –
  Economic evaluation of projects in the electricity supply industry. –
  New ed. – (IET Power and energy series no. 44)
  1. Electric utilities – Finance – Evaluation
  2. Electric utilities – Economic aspects – Evaluation
  I. Title II. Institution of Electrical Engineers
  333.7'9323

ISBN (10 digit) 0 86341 304 8
ISBN (13 digit) 978-0-86341-304-9

Typeset in India by Newgen Imaging Systems (P) Ltd, Chennai
First printed in the UK by MPG Books Ltd, Bodmin, Cornwall
Reprinted in the UK by Lightning Source UK Ltd, Milton Keynes

Contents

Preface

The first edition of this book appeared in 1997, it is now more than five years since then. During that time significant developments (technological as well as managerial) have taken place in the electricity supply industry (ESI) in Europe as well as in many countries in the world. Significant strides in technology and efficiency of power generation have materialised during the past few years as well as advancement in information and telecommunications, which has helped to promote electricity markets and trade.

There is a global trend towards liberalisation and privatisation of the electricity supply industry. This is coupled with growing environmental awareness and increasing prospects for ratification of the Kyoto Protocol. All have dramatically affected the ESI worldwide. Developments that have taken place during the past decade outweight all that has taken place since the middle of the 20th century.

This necessitated a review and a new edition of this work to take into account these developments that affect the way projects are conceived and evaluated by the industry. With the gradual demise of government-owned utilities, financial and economical evaluation of projects is gaining more importance in competitive and liberalised markets, as is risk management.

Financial and economic evaluation of projects is usually carried out by a team of engineers, economists and financial analysts. In the case of economic evaluation of large projects, there is an involvement of economic and environment disciplines, and the undertaking of an analysis that is beyond the proficiency of most engineers, accountants and financial analysts. Projects in the electricity supply industry are slightly more sophisticated than other investment projects in the industrial sector and more capital intensive. It is not easy to introduce people from other disciplines (economists, accountants, financial analysts) into the intricacies of electrical engineering; however, their role remains very important in the evaluation of projects in the power industry. It may be easier to acquaint engineers with the fundamentals of financial and economic evaluation of projects. This will allow for an easier dialogue between different disciplines in any project evaluation team, and correspondingly lead to better results.

This book is written primarily for engineers to assist them in project evaluation; therefore the treatment assumes some understanding of engineering, particularly the electrical power technology and utilisation of computers for simulation. All the financial and economic analyses and examples are brief and simplified to allow for the quick grasp of ideas and understanding of the subject by the non-specialist. However, the book is also useful to economists and financial analysts in understanding issues encountered in engineering projects and the electricity supply industry and the means of evaluating their economic implications. Chapters 1, 8–12 and 14 are, in particular, power engineering biased.

Simplified techniques for the financial and economic evaluation of projects are not difficult to understand by power engineers with their engineering and mathematical knowledge. There are plenty of simple evaluation rules that have to be grasped by engineers and which will give them a good insight to simple financial analysis and economic evaluation that will sharpen their understanding of the economic implications of investments and allow them to understand concepts such as the time value of money, discounting and the discount rate, rate of return and risk evaluation. Such concepts are essential knowledge for power system planners and others who are involved in project selection and strategies.

The book is mainly concerned with the financial and economic evaluation of projects. Therefore it does not dwell on the other related aspects of demand prediction, technology, management and the framework of the power sector planning. These are, of course, very important for financial and economic evaluation; however, they are treated in detail in the available literature, and will be referred to when they are of direct bearing on evaluation. Also this book does not dwell on financial projections of a commercial nature. Therefore the book is mainly concerned with the evaluation of projects, particularly capital-intensive projects like those of the electricity supply industry. Most of the investments carried out by utilities and investors are in the form of individual projects. The evaluation is required to find out the least-cost solution as well as the rate of return on the investment. The whole performance of the investor, as a complex commercial entity, can be assessed through financial ratios and *pro forma* financial statements. These are briefly described in Chapter 6.

Many books have been written in the United States about engineering economics and evaluation of projects, than have been written in the UK. Evaluation is not a universal science; approaches and criteria for assessment differ in the UK and Europe from North American practices in some aspects. The differences are, however, limited and some of them will be pointed out in the book.

International development agencies, particularly the World Bank, the Organisation for Economic Corporation and Development (OECD), different United Nations agencies and other national and regional development funds significantly contributed towards understanding project evaluation, especially the economics of externalities, shadow pricing and environmental damage costing aspects. This book draws on their work wherever relevant.

This second edition is significantly enlarged compared with the first edition. There are three major significant additions. The first concerns environmental considerations and prospects for emissions trading in a carbon constrained world (Chapters 8 and 9), the other is electricity trading and financial management of risk in growing liberalised markets (Chapters 11 and 14) as well as the evolution of the electricity sector and utility of the future (Chapter 12). The second edition contains four additional chapters to cater for these recent and possible future developments in the electricity supply industry.

This book will also be useful for post-graduate electrical power engineering students preparing for a higher degree. It introduces them to the concepts of financial and economic evaluation and the criteria that govern investment in the electricity supply industry.

This book covers a wide ground and could not be written without benefiting from the published effort of others as well as on a lot of literature; therefore every chapter has a list of relevant references, upon which I have drawn, and that will enable the interested reader to sharpen his knowledge about the subject.

Acknowledgements

The analysis in this book has greatly benefited from my extensive experience in project evaluation which I gained while associated with the electricity supply industry world wide as well as in Jordan, through my association with the World Electricity Council and the Arab Fund for Economic and Social Development, a leading regional development institution located in Kuwait. However, any omissions are my responsibility.

I am particularly thankful to the management of the Arab Fund for supporting this work.

Chapter 1

Global electrical power planning, investments and projects

1.1 The value of electricity

Electricity is versatile, clean to use, easy to distribute and supreme to control. Just as important, it is now established that electricity has better productivity in many applications than most other energy forms [1]. All this led to the wider utilisation of electricity and its replacement of other forms of energy in many uses. Demand for electricity is now growing globally at a rate higher than that of economic growth and, in many countries, at almost 1.5–2 times that of demand for primary energy sources. With the type of technologies and applications that already exist, there is nothing to stop electricity's advancement and it assuming a higher share of the energy market. Saturation of electricity use is not yet in sight, even in advanced economies where electricity production claims more than half of the primary energy use. Other than for the transport sector, electricity can satisfy most human energy requirements. It is expected that, by the middle of the 21st century, almost 70 per cent of energy needs in some industrialised countries will be satisfied by electricity [2].

Electricity has become an important ingredient in human life; it is essential for modern living and business. Its interruption can incur major losses and create havoc in major cities and urban centres. Its disruption, even if transient, may cause tremendous inconvenience.

Therefore, continuity of electricity supply is essential. Also, with the widespread use of computers and other voltage- and frequency-sensitive electronic equipment, the importance of the quality of supply has become evident. A significant proportion of investment, in the electricity supply industry (ESI), goes into the reserve generating plant, standby equipment and other redundant facilities needed to ensure the continuity and high quality of the supply. Economics of reliability in the electricity supply industry is a very important topic [3], which will be dealt with in detail in Chapter 10. Optimisation of investment in the ESI requires understanding of markets, prediction of future demand and an approach based on integrated resource planning.

1.2   Integrated resource planning

Integrated resource planning (IRP) seeks to identify the mix of resources that can best meet the future electricity-service needs of consumers, the economy and society. It is the preferred planning process for electric utilities. Under IRP different resource options (incorporating investment and operational costs) are compared using a discounting process and electrical power is visualised as a sub-sector of the energy sector. In turn, the energy sector is one among many other sectors (agriculture, health, education, transport, etc.) that make up the national economy. However, energy and electricity permeate the whole economy; electricity linkage with all other sectors and the national economy at large is stronger than any other sector. Energy is also a capital-intensive activity that claims financial resources and investments more than other sectors, particularly if the country has no sufficient indigenous energy resources [4].

The electricity sub-sector is usually the largest within the energy sector. It claims more than one half of the total capital investment in the energy sector and, on average, consumes almost one third of the country's total fuel consumption (ratios vary from one country to another). Capital investments in the electricity sector of developing countries amounted to almost 9 per cent of their gross capital formation (total capital investment) in the past two decades. This ratio is increasing owing to the rapid growth of electricity demand in DCs [5]. Besides its direct linkages to almost all other sectors of the national economy, the electricity sub-sector also has important environmental impacts of a national, regional and global nature.

Investments in the electricity sub-sector are prompted by the requirements of the economic and social sectors and by demography. However, their extent is restricted by the country's financial and capital investment capabilities and the availability of other energy sources, as well as the continuous improvement in efficiency of utilisation. The form and location of electricity production facilities and fuels utilised are being increasingly influenced by environmental considerations. Within the electricity sub-sector itself, limited financial resources have to be optimally distributed between generation extension and network strengthening, between urban and rural areas and between different geographic locations. Electrical power production also has to utilise different types of fuel and other resources depending on their local availability, cost, and environmental impact. All this calls for integrated electricity planning in a hierarchical manner (see Figure 1.1).

Integrated resource planning in electric power utilities also involves links with the labour, capital and energy markets. It aims at ensuring that the policies (pricing and accessibility to supply) and practices (quality and availability of supply) of the power utilities adequately serve the purposes of the national economy. These purposes include the availability of a reliable supply at the least possible cost, while efficiently utilising resources and protecting the environment.

Electricity is a secondary form of energy. It is produced, in most cases, by converting primary energy into heat that drives turbine-generators, or internal combustion engine-generators. It is also produced directly by hydro-power driving hydro-turbines. Therefore, electricity is an integral part of the energy scene. Fuels for producing

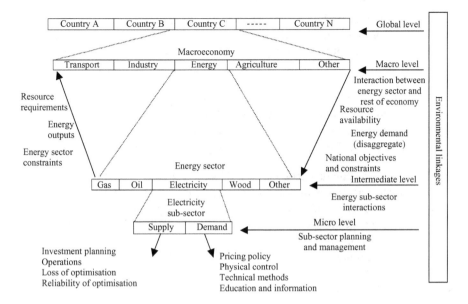

Figure 1.1   Integrated resource planning in the ESI

Integrated resource planning in the ESI seeks to identify the mix of resources that can best meet the future electricity–service needs of consumers, the economy and society. It is presented in the hierarchical framework shown here. Energy is one of the national economy's most extensive sectors and it permeates the whole economy and has linkages practically to all sectors of the economy. Environmental aspects are increasingly linking the electricity sector to the energy sector and through it to the national microeconomy and global environmental management.

Source: IAEA, References [6] and [7].

electricity are the main part of the primary fuels that constitute the Total Primary Energy Requirements (TPER) in any country. Also, electricity can substitute other fuels within the energy sector. Any study into integrated resources management in the electricity sub-sector must consider this relationship. The aim is to have clean production and efficient resources utilisation in the electricity sub-sector. This will not only help to reduce the burden of energy on the national economy, but will also allow efficient and clean replacement of other fuels, thus greatly enhancing integrated resource management within the entire energy sector.

1.3   The changing electrical power industry scene

The electricity supply industry worldwide has resisted rapid change. This is mainly because of the inertia of its large size and investments, and being monopolistic in most cases. Until recently, change has been slow technologically and managerially. Beside the widespread utilisation of semi-conductors and electronics in almost every electrical power facility, technological change has been gradual. Most electricity utilities continued to be monopolies with, in most cases, government ownership and control.

The oil crises of the 1970s shocked the industry and shifted emphasis from expansion into conservation and efficiency. From the late 1980s and 1990s onwards tremendous developments in the management, ownership, and control of the ESI began to take place. These changes were prompted by three important factors.

• The influence of market economies, and its emphasis on competition, which lead to restructuring and deregulation, liberalisation, private sector investment and ownership as well as electricity trade. These developments were made possible by the recent spectacular technological advancement in information technology and communications.
• Rapid changes in technology have occurred in both the generation of electricity and in computing systems used to meter and dispatch power [8].
• Growing environmental concerns that are prompting efficiency, conservation and conversion into cleaner fuels as well as emission-free renewable and distributed generation. Such environmental concerns have advanced international agreements and protocols, the most important of which is the Kyoto Protocol, which will have lasting effects on the future of the ESI [9] (see Chapter 8).

These three factors, which were practically introduced in the last few years and are still gaining momentum, have caused tremendous changes in the ESI. The structural change and technological development that have taken place in the industry in the past ten years outweigh developments since the middle of the 20th century.

The market economy also has led to greater emphasis on the efficiency of investment and profitability of capital. It has encouraged competition, the partnership of the private sector and independent power producers in the industry. Environmental concern encouraged more efficient electricity production, reduction of losses, utilisation of cleaner fuels, abatement of emissions, and control of pollution. In addition, it greatly affected the fortunes of nuclear power. Emphasis shifted from growth into demand management, from more sales into rationalisation of demand, containment of emissions, and better consumer services.

These recent developments are analysed below in greater detail because of their great relevance to the financial and economical evaluation of electrical power projects.

1.3.1   Reform trends in the electricity supply industry

Energy markets worldwide are currently in the midst of a fundamental transformation [10], as a result of technological change and policy reforms. The objectives of these reforms are: to enhance efficiency, to lower costs, to increase customer choice, to mobilise private investment, and to consolidate public finances. The mutually reinforcing policy instruments to achieve these objectives are the introduction of competition (often supported by regulation) and the introduction of private participation. As a large number of developed and developing countries have successfully restructured their electricity and gas markets, an international 'best practice' for the design of the legal, regulatory, and institutional sector framework has emerged. It includes [8]:

- the corporatisation and restructuring of state-owned energy utilities;
- the separation of regulatory and operational functions, the creation of a coherent regulatory framework, and the establishment of an independent regulator to protect consumer interests and promote competition;
- the vertical unbundling of the electricity industry into generation, transmission, distribution, and trade;
- the introduction of competition in generation and trade and the regulation of monopolistic activities in transmission and distribution;
- the promotion of private participation in investment and management through privatisation, concessions, and new entry; and
- the reduction of subsidies and tariff-rebalancing in order to bring prices in line with costs and to reduce market distortions.

These trends greatly widened the scope for financial management in the ESI, introducing to the industry new financial services like risk analysis and risk mitigation, electronic trading, etc. (see Chapters 11 and 12).

1.3.2   Environmental concerns and the efficiency of generating plant

Recently, there has been growing environmental concern about the role of energy, particularly electricity production, in causing emissions that have local, regional and possibly global effects, with long-term detrimental implications. This has prompted emphasis on efficiency as a very important way of curbing emissions.

Electricity generation is an inefficient conversion process. The bulk of the existing inventory of generating plant, globally, has a gross generation efficiency of less than 33 per cent. Vintage coal firing power stations have an efficiency that averages 25 per cent. Significant improvements in efficiency have been achieved during recent years, almost to the extent of 0.2 per cent points per annum. Gross global generation efficiency, over 33 per cent now, was estimated to be around 31 per cent in the 1980s, 29 per cent in the 1970s and as low as 27 per cent in the 1960s (Figure 1.2).

Modern, large and efficient thermal generating facilities utilising pulverised coal have an efficiency approximating 43–44 per cent. Modern combined cycle gas turbine (CCGT) plants claim an efficiency nearing 60 per cent, and in the relatively few instances of combined heat and power (CHP) arrangements, efficiencies of fuel utilisation can exceed 80 per cent. However, owing to the very large inventory of existing inefficient generating facilities worldwide, the average efficiency of electricity generation will not significantly improve in the short term. When the system losses of 10 per cent in industrialised countries, and more than 25 per cent in many developing countries, are taken into account, the amount of energy reaching consumers in the

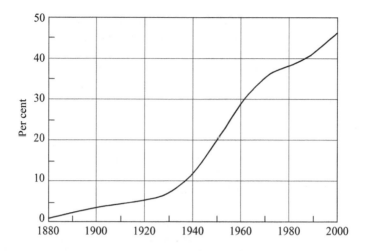

Figure 1.2   Development of electricity generation efficiency in thermal plants utilising pulverised coal [11]

form of electricity is less than 30 per cent of the calorific value of the fuel infeed. In some developing countries it may be even lower than 20 per cent.

1.3.3   Environmental concerns and the growing importance of natural gas

The growing importance of natural gas (also liquefied natural gas, LNG) as a source of energy, particularly for electricity production, has enhanced the utilisation and development of the gas turbine and its derivative (the high-efficiency combined cycle plant) at the expense of the capital-intensive traditional steam power stations.

Natural gas was a premium fuel in the past. Recently, however, owing to its rapidly increasing reserves, abundance, relative cheapness and cleanliness, it has become the fuel of choice for electric power generation, whenever available. Environmental considerations have played a considerable part in this regard. Compared with other fuels, natural gas is a benign fuel. It has practically no sulphur, correspondingly no sulphur dioxide (SO_2), and it emits no particulate matter (PM). Its emissions of carbon dioxide (CO_2), which is the main greenhouse gas, and nitrogen oxide (NO_x) are almost half those of coal. This is detailed in Table 1.1.

Increasingly, more global natural gas production is used for electricity generation. Utilisation of natural gas and LNG for power generation will gradually grow more than that of any other fuel. However, its usage will continue to be limited by its availability in only a few countries, its high cost of transmission through pipes, the need for long-term contracts and the high cost of LNG. Only one fifth of the present production of natural gas is traded internationally. Natural gas projects cost more than twice the amount of crude oil projects. Its transport may be six times as expensive. All these factors are slowing the development and utilisation of natural gas as the ideal fuel for electricity production.

Table 1.1   Emissions from fossil fuels [12]

Fuel	Air pollution emission (million tons/m.t.o.e. of fuel)			
	SO_2	NO_x	CO_2	PM
Coal (3% S)	0.081	0.018	3.57	0.106
Coal (1% S)	0.027	0.018	3.57	0.096
Fuel oil (residual)	0.060	0.017	3.13	0.004
Fuel oil (distillate)	0.006	0.009	3.19	0.002
Natural gas	—	0.012	2.07	—

Note: m.t.o.e. = million tons of oil equivalent.

1.3.4   Rehabilitating, retrofitting and repowering of existing power facilities [13]

Stringent regulations (environmental and otherwise) and waylaying conditions, which increased the cost of new facilities, led to frenzied attempts to rehabilitate, refurbish and repower existing facilities. Shortages of capital also contributed towards extending the life and improved utilisation of existing sites. Difficulty in licensing new nuclear facilities led to extending the life of existing facilities to 50 years.

Repowering involves modification of an existing plant by changing the method used to generate power. Repowering may modernise the power plant, increase its efficiency, lower operating costs and/or increase output. Evaluation of the economics of repowering and retrofitting existing sites needs more sophisticated evaluation than other projects. This will be detailed in later chapters.

Markets, particularly the EU energy market, require environmentally compliant and economically competitive plants. Some power plants in Central and East Europe are generally ageing, not well maintained, have high pollution and low efficiency. Their technology is outdated. Consumers are direct victims of this ageing infrastructure as they pay a high cost for power.

The business options for an investor/owner of generation in the EU to meet the new pollution limits include demolishing and rebuilding, retrofitting with emission-control equipment or repowering to make use of low-emission gas-fired CCGT technologies. Demolition and replacement with new efficient equipment requires high capital costs. Retrofitting appears to be a valid option because of the high EU market environmental requirements. If new pollution-control equipment can be retrofitted for a relatively low cost and the plant is still reasonably competitive, then this option should be considered. However, retrofitting is limited to the installation of modern pollution control devices in order to meet emission limits. It does not improve efficiency or reduce the cost of power production.

If the existing plant is not competitive, retrofitting will not help, and repowering may be a viable option. Repowering, that is making substantial changes to power plants at the end of their useful economic or environmental life, improves efficiency,

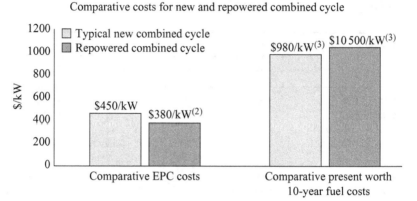

(1) Assumes no opportunity cost for repowered facilities.
(2) 540/kW saving can be assumed that it is directly attributable with avoided steam turbine costs. Total savings of $70/kW may be due to reuse of existing cooling system, water treatment, and other site facilities. Specific cost estimates may differ significantly from these estimates.
(3) A 3 per cent gas escalation and 10 per cent discount rate is assumed.

Figure 1.3   Economics of repowering generating plant [13]

decreases environmental emissions and reduces operating cost. If existing equipment can be reused, such as the steam turbines, repowering may be cheaper than rebuilding. Complete replacement of major systems such as boilers and turbines, conversion to combined-cycle operation (through the addition of combustion turbines and heat recovery equipment) or introduction of circulating fluidised bed (CFB) boilers are the options to be considered for ageing power plants to comply with the EU standard (see Figure 1.3).

For repowering to be economically attractive, owners must look for repowering returns equal to their expected returns on other options. Depending on the state of the facility, income and investment vary and so do the returns:

- if the existing facility is physically or economically forced out of service, 'as is' net operating income is zero;
- if the facility is operable and capable of cycling duty and sales into a competitive market, the 'as is' income may be surprisingly high; and
- if the facility requires extensive investment (retrofit) for environmental compliance, 'as is' investment may be quite high.

The capital investment required to repower ageing plants is offset by the life extension of the plant, the decrease in operations and maintenance costs (O&M) and increased efficiency and power generation. Repowering with natural gas turbines can cost effective compared with building a same size new combined-cycle unit, because existing equipment may be re-used.

1.3.5   *The growing importance of demand side management (DSM)*

Prompted by the high cost and shortage of capital, many power utilities have found it cheaper and more profitable to invest in demand side management (DSM) rather than extending the supply and building new facilities [14].

DSM activities vary and involve the following.

- Changing the shape of the daily and seasonal peak load curves (reducing peaks and filling the load curve valley).
- Introducing and encouraging the use of more efficient electric apparatus.
- Reducing waste and overuse, through pricing, regulations, and educating the consumers. Under this category falls reducing system loss.
- Conservation: beyond eliminating overuse, conservation helps to manage and reduce demand without affecting the quality of life.
- Substitution: some electricity benefits can be substituted more economically by other means. For instance, passive solar design of buildings can significantly reduce demand for lighting, electric and water heating.

Methods of evaluation of the benefits and economics of DSM projects are referred to later.

Most of these considerations will have beneficial effects on the electricity supply industry. They can lead to cleaner and leaner production facilities, more rational and efficient use as well as fostering competition that assists in optimising the use of resources; thus lowering costs and giving greater benefits to the consumer. Investment requirements will be rationalised and electricity costs can be reduced. All this, however, is going to be gradual and will vary from one country to another. The ESI is different to telecommunications. Energy systems, particularly electricity systems, have huge inertia; they are highly capital intensive and live for a long time. They also demand a lot of licenses and way leaving. This, of course, delays reaping the beneficial results of some of these recent developments. Such delays have also been assisted by slow technological change in the way electricity was being produced, transmitted, or distributed in the past. New and renewable energy sources (other than hydro) have not yet lived up to their earlier promise or our expectations. Such sources, which can be mostly only utilised as electricity, will take a dramatic boost in the future, if the present worries about global warming are reinforced by more substantive arguments or proof, and once the Kyoto Protocol is ratified. However, liberalisation of the markets is creating a new potential for the power industry. Technological change will increasingly affect the way electricity is being generated and distributed. Distributed generation, fuel cells, the virtual power plant and similar technological innovation assisted by market liberalisation are gradually, but surely, changing the power industry into a new future.

1.4   The global electrical power scene

Total global electrical power production in 2001 amounted to around 15 700 terawatt-hours (TWh), with an average of 2600 kilowatt-hours (kWh) per capita per annum.

The global installed electric power facilities are very difficult to quantify exactly. Estimates vary from 3430 GW [15] in 2000 to over 4200 GW [16], with the prospect of the latter figure being more accurate. Most of the world electricity production, 57 per cent, was in OECD countries with an annual average of 8800 kWh per capita; and another 10 per cent in East Europe and the former Soviet Union. Developing countries, which account for more than three quarters of the world's population, utilised only 33 per cent of the global electricity production, with an annual average of 1100 kWh per capita.

Thermal power, mostly utilising solid fuels, accounted for 64 per cent of global electricity production. Hydro-produced power accounted for 18 per cent and nuclear for 16 per cent. Total fuel utilised to produce this electricity amounted to around 37 per cent of the world's primary commercial fuel use. Detailed statistics of electrical power in the three economic groupings (OECD, East Europe and developing countries), energy, input fuels used and types are shown in Table 1.2.

1.4.1   Electricity and the global energy scene

As previously explained, electricity is part of the energy sector and demand for electricity is part of the total demand for energy. However, because of its virtues and cleanliness, electricity is gradually substituting other forms of energy and enhancing its role in the global energy scene.

Global primary energy consumption in 2001 amounted to over 9800 m.t.o.e. Of this amount, 8900 m.t.o.e. were from commercial energy sources (oil, gas, coal, hydro, nuclear). Non-commercial energy (waste, fuel wood and other forms of biomass) is difficult to estimate, and accounts for the remainder. Of this global energy consumption around 3450 m.t.o.e. were utilised to produce electricity. This amounts to around 37 per cent of the total global commercial energy primary use. In 1950 and 1970 energy input to produce electricity amounted to 18 per cent and 29 per cent of the total primary consumption, respectively [17]. The growing role of electricity in the global energy scene will continue in the future, as previously discussed.

1.4.2   Electricity and the world economy

There has always been a tight coupling between energy use and economic growth. Demand for electricity has always displayed good correlation with that of global economic growth, as measured by gross national product (GNP) in purchasing power parity dollars ($ppp). This strong coupling between electricity and economic growth is clearly shown in Figure 1.4.

During the past 25 years, electricity growth averaged 4 per cent per annum, which is a high rate of growth. In the future, electricity demand growth is expected to match the growth of the world economy. This is expected to average around 2.5–3.5 per cent annually during the next few years.

The International Energy Agency (IEA) and the International Atomic Energy Agency (IAEA) estimate that global electricity production will increase at an annual average rate of 2.7–3 per cent during the first decade of the 21st century [19]. Therefore it is expected that total electricity production in 2010 will amount to around

Table 1.2   Global electricity statistics

	Population (million)	Capacity (GW)	Production (TWh)	Electricity % kWh per capita		Energy (m.t.o.e.) Total primary energy	Electricity m.t.o.e.	Elec. (%)
Global electricity balance (2001)								
OECD	1020	2000	9000	8800	57	5000	1970	39
East Europe	400	420	1600	4000	10	1070	420	39
DCs	4580	1160	5100	1100	33	3430	1060	31
World	6000	3580	15 700	2600	100	9500	3450	37

	GW	TWh	% of electricity	Input (m.t.o.e.)	
Fuels for electricity generation					
Nuclear	360	2500	16	650	18%
Hydro	830	2900	18	250	7%
Thermal:	2330	10 000	64	2560	72%
solids	1100	6100	39	1570	44%
oil	430	1350	9	290	8%
gas	800	2550	16	700	20%
Other	60	300	2	110	3%
Total	3580	15 700		3570	

	Nuclear (TWh)	Hydro (TWh)	Thermal and other (TWh)	Total (TWh)
Sources of electricity generation (2001)				
OECD	2000	1300	5700	9000
East Europe	240	290	1070	1600
DCs	60	1310	3730	5100
World	2500	2900	10 300	15 700

Source: IAEA: Energy, Electricity and Nuclear Power Estimates for the period up to 2020 (July 2001).
IEA: World Energy Outlook (2000).

Notes:
(1) Generation refers to gross generation.
(2) Capacity refers to net capacity.
(3) Fuel required for production of electricity is estimated at 255 g of oil equivalent per kWh for high income and 270 g kWh^{-1} for East Europe and developing countries.
(4) Fuel equivalent of hydro generation is equal to the energy content of the electricity generated.
(5) m.t.o.e. = million tons of oil equivalent.
(6) Terawatt-hour (TWh) = 1000 million kWh, Gigawatt (GW) = million kW.
(7) Global electricity figures are not well documented. The above is a collection of data from many sources with some rounding approximations and estimation.

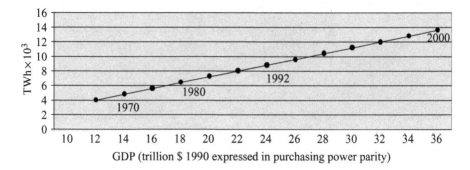

Figure 1.4   Electricity demand as a function of world GNP (excluding former CPEs) [18]

20 000 TWh and in 2020 to 25 880 TWh. Most of this growth is going to occur in developing countries, particularly in south-east Asia, a region that is enjoying a rapid economic growth. In 2030 global electricity production is expected to exceed 28 000 TWh. Half of this amount will be accounted for by developing countries [20].

1.5   Considerations that will influence future investments in electricity generation

Environmental considerations [20] will greatly continue to influence the generating system through choice of technology and fuels, particularly if there is future carbon taxation. Electricity generation is highly capital intensive and the inertia of existing generation systems is considerable. Existing systems involved huge amounts of capital investment, which every electricity producer is eager to use to the end of its useful life. Because of the huge stock of current generating systems, technological change will be very gradual. Choice of fuels will also be restricted by: the existing stock of plants, national fuel endowments (particularly in the case of coal), capital and human resources (in case of nuclear), development of natural gas networks and its future price, and availability and cost of alternative resources. Correspondingly, the effect of technological changes in the electricity generation system will be slow and gradual, at least in the short term, since most of the expansion will take place in developing countries. These countries are more concerned about capital expenditure, utilisation of proven technologies and cheaply available local coal fuel, rather than the global environment and efficiency considerations.

1.5.1   Future electricity generation technologies

Future technologies [10] will aim, particularly in OECD countries, at achieving the following:

(i)   a high efficiency to limit fuel use and emissions,

(ii)   clean emissions to reduce environmental impact,
(iii) low capital cost to limit investments and
(iv)   limited capacities with short lead times and distributed sources to minimise uncertainties and risks.

Modern CCGT firing natural gas, largely, meet most of these conditions. Nuclear, hydro and renewable sources meet some of the conditions. Existing coal technologies, however, are the least capable in honouring them. Therefore, the emphasis is on developing environmentally sound and clean coal technologies, such as that offered by fluidised bed combustion (FBC) and zero carbon emission technologies. Advances have been achieved in improving efficiency of electricity generation, particularly in recent years. However, to reduce CO_2 emissions further, new technologies for gas- and coal-based combined-cycle generation are needed. A few such technologies are currently available and their commercial application lies not far into the future. These are, for example, pressurised fluidised bed combustion (PFBC), integrated gasification combined-cycle (IGCC), and the topping cycle, as well as the zero carbon emission technologies that are presently being developed.

1.5.2   *Future electricity generation fuels*

The major energy sources that currently dominate the generation of electricity (oil, natural gas, coal, hydro, and nuclear) will continue for a few decades. As mentioned earlier, this is because of the slow development of economically new and renewable energy sources, and, of course, the vast existing stocks of facilities. Owing to its high price, crude oil and its products will continue to diminish their share in electricity generation and, most likely, in absolute terms. The main utilisation of crude oil will continue to be in the OPEC, and in small developing countries, which have no local energy sources. Natural gas is an ideal fuel for electricity generation and, correspondingly, it will increase its present share (16 per cent) of the electricity fuels market. The prospects of it assuming a much larger share will be enhanced in the case of future energy/carbon taxation and ratification of global environmental agreements. Its use will, however, continue to be limited by the previously mentioned factors and uncertainties about future gas prices. Coal is the world's major fuel for electricity generation and will remain so for many years to come. Its technologies are well established, prices are cheap, and reserves abundant. Regardless of its pollution problems, it will benefit from the fact that most of the growth in electricity generation will be in developing countries that are relatively rich in coal, like China and India. Developments in technologies for carbon sequestration will greatly enhance the fortunes of coal.

The future of nuclear energy, regardless of its established technologies and absence of gas emission, will continue to be handicapped by factors such as: limited public acceptability, large size of units, and huge investment requirements. It must also be realised that its energy utilisation is limited to electricity production and of the future growth in electricity requirements will mostly be in developing countries that, often, cannot meet the investment and technology requirements. Therefore, for the future, and other than for unpredictable environmental events (surprises), nuclear energy's relative contributions towards global electricity requirements will, most likely, not increase. At best, its present contribution of around 16–17 per cent will be maintained.

Like nuclear energy, hydro-energy can only be utilised for the production of electricity. However, in contrast, hydro-energy has limited economically viable sites, with most of the large sites already utilised. Other viable sites, like in Africa, have no nearby loads. Therefore, its future proportional share of electricity production could decrease. During the 1980s, hydro-energy grew at a handsome 2.3 per cent annually, but this has limited future prospects, since new sites demand higher capital investment and have a greater environmental impact.

The contribution of new and renewable energy sources such as solar, photo-voltaic, wind, geothermal, etc., will be growing very fast in percentage terms but their global utilisation will continue to be limited because of present limited facilities and small unit size [21]. Solar energy, although abundant, is dispersed and ineffi-cient and correspondingly expensive to exploit. This applies to almost all forms of new and renewable energy sources for electricity production; however, wind energy has shown recent economies of scale and solar cells are increasingly being used for special applications. Even if there is a technological breakthrough in the future (which is not yet evident), the effect will be gradual and limited, given the enormous stock of existing generation facilities. Prospects for wind energy and solar cells (PV), which have the best prospect for rapid growth, are detailed in Chapters 8 and 9.

1.6   Projects and capital investments

The electricity supply industry is highly capital intensive. It is probably more capital intensive than any other sector, particularly in the developing countries. Therefore, planning and proper financial and economical evaluation of projects are important to rationalise investment and achieve economic efficiency.

A review of recent studies [5] indicated that there is a need to invest $2700 billion (in 2000 dollars) in the electricity supply industry worldwide over the period 1995–2010, that is, $175 billion on the average per annum; this is equivalent to £120 billion pounds sterling annually. Generation projects will account for 63 per cent of these capital investments, transmission for 8.8 per cent, and distribution for 21.2 per cent. The remaining 7 per cent will be general expenditure, mainly concerned with control, telecommunications, and similar activities. Developing countries will account for over 58 per cent of these investments, developed countries (OECD) for 32 per cent and the Transitional Economies (East Europe and the former Soviet Union) for the remaining 10 per cent, as detailed in Table 1.3.

1.6.1   Generation investment and projects

Most of ESI projects will be in generation, and will take place in developing countries. To the existing stock of global generation capacity of 3580 GW in the year 2000, it is expected that 1200 GW of extra capacity will be added, that is, an increase of at least one third over 15 years. Of this, China alone is expected to account for almost

Table 1.3   Capital needs by use and country type (in billions of 2000 dollars), over the period 1995–2010

Type	Generation	T & D and General	Total	%
Developed	560	300	860	32
Developing	970	615	1585	58
Transitional	170	85	255	10
Total	1700	1000	2700	100

Source: modified from Reference [5].

20 per cent of the new generation capacity, by investing in 230 GW of new capacity over the period. This is more than the whole extension of capacity of Western Europe that is supposed to increase by 200 GW. Costs per MW of capacity vary from one country to another, depending on the type of generation, fuels utilised, and more importantly investments in environmental protection facilities.

For most of the new generating facilities, it is expected that 40 per cent will fire natural gas or LNG, 35 per cent coal-fired capacity, and 20 per cent will be new hydro-electric plants. The remaining 5 per cent can be in nuclear and new and renewable electricity generating facilities. However, coal-fired generating plants, because of their capital intensive nature, are expected to account for more than half of the generation capital investment. Generally, the 1200 GW of new generating plants (with all environmental facilities) are expected to cost $1700 billion, which is an average of $1400 kW^{-1}, i.e. £860 kW^{-1} in 2000 money.

Coal is expected to continue to be a favoured fuel for new power production projects in many developing countries, since most of the new generating facilities will be in south Asia and China which are rich in local coal resources.

Gas-fired capacity is expected to grow proportionally faster than any other type. Such facilities will increasingly assume the role of base load generation in almost all regions of the world, except in south Asia and China, which have very little natural gas reserves. Natural gas prospects can greatly be enhanced by the growing environmental awareness. Hydro-electric power generation is also expected to grow in China and Latin America. However, in some African countries, demand shortage delays utilisation of the abundant hydro-power capacity, while in OECD countries most of the economically exploitable sites have already been utilised.

Future nuclear power projects, over the same period 1995–2010, are expected to be few (70–80 GW), owing to the very high cost of nuclear facilities, long lead times, and public awareness. Most of the new facilities will be in Japan, China, east Asia and south Asia, and in Finland.

1.6.2   Transmission and distribution projects

As shown by Table 1.3, transmission, distribution, and general investments are expected to amount to $1000 billion (£660 billion) over the 1995–2010 period.

Most transmission projects will take place in developing countries, particularly Latin America, where large transmission lines will be needed to convey electricity from remote hydro-electric stations. A few transmission facilities will be needed in Europe, where the network is well developed. Deregulation, unbundling of the transmission network and difficulties in getting way leaves are creating bottle necks in transmission facilities in many developed countries.

Most developed and developing countries need to invest sufficiently to extend and improve their distribution networks. Losses in the distribution system include most system losses in the developing countries. Projects to reduce losses and improve the supply continuity will figure highly in developing countries' distribution plans. Therefore, it is expected that these countries will invest more than twice as much on their distribution facilities compared with transmission. Strengthening and extension of rural networks, in many developing countries, will be the centre of interest. At the time of writing there are some two billion people, mostly living in rural areas, who have no accessibility to electricity supply. Providing them with electricity is the greatest challenge facing developing countries in their pursuit of sustainable development.

1.6.3   Sources of investment funds

In 2000, the world's Gross National Income (GNI) was around $31 171 billion [22] (in purchasing power parity it was around $44 500 billion). It is expected to increase to around $42 000 billion (2000 dollar) in 2010. With investments averaging 20 per cent of this GNI globally, the investments in the electricity supply industry will average 3–4 per cent of the world's total investments (gross capital formation) over the next few years. As for the developing countries, this ratio is almost double; the investments in the electricity supply industry will average 8–9 per cent of their total investments. This is a heavy burden on the economies of most countries in the world, particularly the developing countries.

In the past, self-financing, as well as governments and development funds and agencies, used to provide the required funds for investment in the electrical power system, and this is still the case in many developing countries. Owing to privatisation and restructuring, markets are being called upon to provide most of the needed investments in developed countries and, slowly but gradually, in developing countries. Investments in the electricity supply industry in OECD countries do not exceed 1.3 per cent of their gross domestic investment (GDI). Therefore, they represent no practical problem. Developing countries, however, need to invest $1585 billion in their electric power systems during the period 1995–2010. Because of subsidies and low electricity tariffs, internal fund generation in these countries is limited. Government capabilities to finance such infrastructural projects are becoming more limited, owing to the increasing demand of the other more pressing social sectors (education, health, poverty eradication, nutrition, etc.).

Official Development Assistance (ODA) to developing countries has decreased during recent years. Therefore, the financing gap of electrical power projects is increasingly being filled by calling on foreign private capital through attracting investment and the financial markets.

Table 1.4   Investment in infrastructure projects with private sector participation in developing countries by sector and region 1990–1999 (billions of $)

Sector		Region	
Telecommunications	249	Latin America	286
Energy	193	East Asia	169
Transport	106	EE & Central Asia	63
Water and Sanitation	31	Middle East and Other	61
Total	579	Total	579

Source: Reference [22].

Recent studies by the World Bank have shown that, over the past decade, the telecommunication sector attracted more private participation in developing countries followed by the energy/electricity sector, as detailed in Table 1.4. Most of these investments went into Latin America. However, total private participation in energy projects (including oil and gas) over the 1990–1999 period of $193 billion is small compared with developing countries' requirements of over $1585 billion over the 1995–2010 period.

It is not possible to ascertain how much of the above-mentioned private capital is going into power development in developing countries. However, the electricity supply industry attracts a sizeable amount of this capital and an increasing proportion will have to be provided by international private capital. For this to be achieved, much reform (privatisation and restructuring) and regulatory arrangements must take place in developing countries. Three conditions are mandatory [23].

- Governments must be committed and guarantee a financially independent electricity supply industry.
- Electric utilities have to perform in a financially and economically viable way.
- Investors (foreign as well as local) must be convinced that they will obtain a good return on their investment, and will be able to repatriate these returns.

Correspondingly, the financial and economical evaluation of projects in the electricity supply industry, and the techniques for the calculation of return requirements, are becoming increasingly important.

1.7   Epilogue

This book is written at a time of major change, which will reshape the future of electricity, owing to the following considerations [23,24].

(i)   Global environmental concerns, particularly regarding detrimental effects of carbon emissions, are paramount. OECD countries, and increasingly other countries, are seriously considering these concerns when making investment decisions. With more environmental knowledge in the future, things may become even tighter. However, this developing environmental awareness is promoting demand management and an efficiency drive that will greatly influence electricity's future.

(ii)   There is uncertainty about the future of nuclear energy.

(iii) The transformation into market economies is now taking place, not only in Central and Eastern Europe, but also in a growing number of developing countries. This transformation, from central to market planning, will greatly influence the future electricity and energy use of these countries.

(iv)   With the advancement in information and telecommunication, electricity trade (locally and regionally) is becoming common. Electricity markets, similar to financial markets, are growing.

(v)   There is a growing trend towards deregulation and privatisation of the electricity sector in major electricity consuming countries, with significant implications of this on investment decisions and demand patterns.

(vi)   There are advances in technology and efficiency, particularly in medium-sized, combined-cycle generating facilities.

(vii) There is a significant level of capital shortages in most developing countries.

All the above factors are creating uncertainty in the electricity supply industry, significantly affecting investment decisions, and sometimes delaying them.

This calls for a more thorough evaluation of the financial and economic impact of projects in the electricity supply industry. There is a need to move from deterministic assumptions, for studying and modelling of future plans, into risk analysis and mitigation and valuation of uncertainties. Least-cost planning based on deterministic assumptions, does not explicitly evaluate economic costs and risks of failure to achieve economic objectives. Therefore, the results of the investment are suboptimal. Also, deterministic methods do not help in understanding the mechanisms that greatly influence investment decisions, like the discount rate and impact of environmental factors. Such aspects will be dealt with in greater detail in the coming chapters.

1.8   References

1 JARET, P.: 'Electricity for Increasing energy efficiency', *EPRI J.*, April 1992, **17**, (3)

2 GERHOLM, T. R.: 'Electricity in Sweden – Forecast to the year 2050' (Vattenfal, Sweden, 1991)

3 KHATIB, H.: 'Economics of reliability in electrical power systems' (Technicopy, Glos, 1978)

4 NAKICENOVIC, GRUBLER, and MCDONALD: 'Global Energy Perspectives' (Cambridge University Press, 1998)

5 'Financing worldwide electric powers, can capital markets do the job?'. Resource Dynamic Corporation for US Department of Energy, April 19, 1996

6 'Senior Expert Symposium on Electricity and the Environment', International Atomic Energy Agency (IAEA), May 1991, Helsinki, Finland

7 MUNASINGHE, M.: 'Energy Analysis and Policy' (Butterworth, London, 1990)

8 BACON, R. W. and BESANT-JONES, J.: 'Global Electric Power Reform, Privatisation and Liberalisation of the Electric Power Industry in Developing Countries', Annual Reviews – Energy and the Environment. *The World Bank*, 2001

9 'Siemens: Pictures of the Future – Emission Certificate Trading', Siemens Picture of the Future, Spring 2002

10 MINCHENER, A.: 'Coal comes clean', *IEE Rev.* Nov./Dec. 1991, **37**, (11)

11 SIMON, M.: 'The road to environmentally compatible power and heat generation', *ABB Rev.*, **3**, (91)

12 'Greenhouse gas emissions – the energy dimension', OECD/IEA, Paris, 1991

13 BLARK and VEATCH: 'An Attractive Option', *Electricity International*, May 2002, pp 26–28

14 NADEL, S. and GELLER, H.: 'Utility DSM', *Energy Policy*, April 1996, **24**, (4)

15 'Energy, Electricity and Nuclear Power Estimates for the period up to 2020', IAEA, Vienna, July 2001 edition

16 'World Electric Power Plants' (Utility Data Institute, Division of the McGraw Hill, 2nd edn., and later editions, 1996)

17 KHATIB, H. and MUNASINGHE, M.: 'Electricity, the environment and sustainable world development', World Energy Council 15th congress, Madrid, September 1992

18 BOURDAIRE, J.-M.: 'The link between GNP and energy consumption', Oxford Energy Forum, May 1995, **21**

19 '1996 World Energy Outlook', IEA, Paris, 2000

20 KHATIB, H.: 'Electricity in the global energy scene', *IEE Proc. A*, June 1993, **140**, (1)

21 ARGIRI, M. and BIROL F.: 'Renewable Energy', Oxford Energy Forum, Issue 49, May 2002, pp 3–5

22 'World Development Report 2002', The World Bank, Washington DC, 2002

23 MULLER-JENTSCH, D.: 'The Development of Electricity Markets in the Euro-Mediterranean Area', World Bank Technical Paper No 491, Washington DC, 2001

24 'Energy and the Challenge of Sustainability – World Energy Assessment', UNDP, UNDESA, WEC, New York, September 2000

Chapter 2

Considerations in project evaluation

Projects involve investments that are meant to satisfy a demand, and to achieve an engineering or economic purpose (usually better efficiency or enhanced perform-ance). Therefore, a project is a process of creating specified results. It is a complex effort involving many tasks, to achieve a certain objective. A project is a non-repetitive unique process with start and end points, budgets and financial plans, life-cycle phases, and stages. Projects can be capital intensive (electrical power, energy, telecommuni-cations), infrastructural, civil work intensive (transportation, water supply, etc.), and people intensive (agriculture, education, nutrition, etc.) [1,2].

2.1 Project selection and evaluation

Project selection is a problem of allocation of scarce resources including capital, skilled labour, management and administrative capacity, as well as other resources (land, energy, etc.). In the electricity supply industry, most projects are rather imposed on the industry by rising demand. However, there are many ways of sat-isfying a demand, through building new generation facilities, repowering existing facilities, strengthening the network, or rationalising demand by demand side man-agement (DSM) plans. Many projects are prompted by the need to improve the quality of supply to achieve better continuity or adequate system standards. Some projects, for efficiency improvement, are justified by economic and environmental considerations.

Therefore, the requirements for projects in the electric power sub-sector are more focused and less diverse than those encountered in other sectors. Recently, with deregulation and privatisation, a wider market has become available for entrepreneurs and investors to sponsor projects, mainly generation projects in electricity utilities. With increased globalisation, investor funds in the electricity supply industry will be seeking global as well as local markets. Although some projects undertaken by inde-pendent power producers (IPPs) have enough financial and demand guarantees, like take-or-pay clauses, they still contain an element of risk – project cost, delays, plant availability, etc. Therefore, project selection, analysis and evaluation are becoming more important in the electricity supply industry than at any time in the past.

Project analysis is a method of presenting the choice between competing uses of resources, and is done through analysis of information and data. Project evaluation studies are meant to assist in the selection and design of new viable projects. A study will evaluate the extent to which the project produces the intended results, the proper technology, the least-cost alternative process, as well as the cost-effectiveness of the project. It will also consider the engineering as well as the financial risk and evaluate the economic (social) cost.

2.2   Project development

The development of projects is a cycle involving three distinct phases: the pre-investment, investment, and operational phases. Capital-intensive projects in the electricity supply industry pass through these three phases with each of the first two phases divided into stages of planning, design engineering, and execution stages [3,4].

2.2.1   The pre-investment stage

The pre-investment stage begins with the identification of the need for the project (demand that has to be satisfied by new generation facilities, or a bulk supply sub-station, or an investment opportunity from an independent power supplier). This preliminary identification is followed by a pre-feasibility study, which is viewed as an intermediate stage between the identification of the project and a detailed feasibility study. In the pre-feasibility study, a detailed review of the need for the project and demand for its output is undertaken, as well as of the possibilities and alternatives (site, size, fuels, etc.). Therefore, at this phase, a lot of support studies are undertaken: market and demand studies, fuel supply, cooling water, location studies including soil mechanics, environmental impact assessment, as well as financial and economic analysis into the selection of the least-cost facilities, technologies and location, and profitability of the project.

The feasibility study

The feasibility study should provide all data and details necessary to take a decision to invest in the project. The feasibility study defines and critically examines the results of the studies undertaken in the pre-feasibility stage (demand, technical, financial, economic and environmental). The results of the feasibility stage are a project where all the background features have been well defined: size and location of the facilities, technical details, fuels, network features, environmental impact and how to deal with it, timing of the project, and implementation schedule. The financial and economical part of the feasibility study will cover the required investment and its sources, the expected financial and economical costs and returns.

It is not possible to draw a clear dividing line between the pre-feasibility and the feasibility studies. The pre-feasibility study is concerned with defining the alternatives (demand, location, size, technology, fuels, costs and environmental impact), while the feasibility study defines the project in a manner that allows implementation to proceed. A typical outline of a feasibility study is shown in Table 2.1.

The feasibility study for large electrical projects (like building a new large power station or a major bulk transmission network and substations), has not only to be confined to the project, but has to look into the electrical power sector in which the project will be operating (through undertaking power system analysis), the future demand, and market for electrical power. The feasibility study has to undertake an engineering analysis that looks into the technologies, the scope of the project, its timing and implementation arrangements, its management and manpower requirements, and its financial and economic viability.

In this book, we are mainly concerned with the financial and economic evaluation of projects to ensure their viability. However, it is essential to understand the impact of the other four considerations in electrical power project feasibility:

(1)   the sector;
(2)   the market (demand);
(3)   technology and engineering analysis;
(4)   management and manpower.

These are briefly dealt with below, to introduce the project evaluator to the different aspects that need to be considered before undertaking a project. In such capital-intensive projects, the financial plan is of vital importance; it has to ensure that there are enough funds to carry out the project into completion. Funds are needed not only to finance the project cost, but also to provide the working capital and pay for interest during construction, and for ancillary costs and expenses. In most such expensive projects, beside the capital (equity) from the project owner, there are loans provided by banks, financial institutions, and development agencies. All of these needs to be investigated, arranged and coordinated in the financial plan.

Project appraisal

Apart from the project pre-feasibility and feasibility studies that are carried out by the project owners, more evaluation, which is termed 'project appraisal', is executed by the financiers (other than equity owners like lending banks or development funds). The idea of appraisal is to satisfy the financiers as to the accuracy and soundness of the feasibility study. Appraisal usually deals in depth with macro-economic matters, environmental impact, and externalities. These are costs and benefits that are outside the confines of the project itself, but affect other objectives or policies of the country. Appraisal work is greatly assisted by having a properly prepared feasibility study that covers every facet of the project. However, appraisal goes beyond feasibility into considering policies, regulations, and other macro-economic considerations and externalities. Appraisal will undertake a thorough economical (social) evaluation of the project costs and benefits, as well as carry out detailed sensitivity and risk analysis; this is to ensure the financial and economic viability of the project.

Table 2.1   Typical outline of a feasibility study [3]

---

1. Introduction
2. The sectoral setting
    (a) The industrial (electrical power engineering) sector and linkages to the rest of the economy
    (b) The sub sector (e.g. the generation sub sector)
    (c) Issues and problems
    (d) Proposals for change
3. The market, pricing and distribution
    (a) The market
        (i)   Historic supply and consumption
        (ii)   Projected demand and supply
        (iii) Market for the proposed project
    (b) Transmission, distribution and marketing
    (c) Pricing
4. The utility
    (a) Background
    (b) Ownership
    (c) Organisational framework
    (d) Management
5. The project
    (a) Objectives
    (b) Scope of the project
    (c) Technical description
        (i)   Production facilities
        (ii)   Utilities and infrastructure
        (iii) Ecology and the environment
    (d) Manpower and training
    (e) Major inputs
    (f) Project management and execution
    (g) Project timing
    (h) Environmental impact and measures for environmental preservation
6. Capital cost and financing plan
    (a) Capital cost
    (b) Working capital requirements
    (c) Financing plan
    (d) Procurement
    (e) Allocation of financing and disbursement
7. Financial analysis
    (a) Revenues
    (b) Operating costs
    (c) Financial projections
    (d) Break-even analysis

Table 2.1   Continued

    (e)   Accounting and auditing requirements
    (f)   Financial rate of return
    (g)   Major risks and risk analysis

8. Economic justification
    (a)   Economic analysis and economic rate of return
    (b)   Linkages and employment
    (c)   Technology development and transfer
    (d)   Foreign exchange availability and effects
    (e)   Regional development impact

9. Agreements

The project appraisal, particularly when carried out by regional international development agencies, leads to a thorough critical evaluation of the project and the sector. It often comes out with suggestions and proposals that improve the project's setup and enhance the future performance of the sector. Some of these proposals are crucial to the success of the project, so that they are treated as covenants and are incorporated in the financing and loan agreement, and the effectiveness of the loan agreement and its disbursement is conditional on the prior honouring of such covenants. Therefore appraisal is not only beneficial for the project but can also lead to sector reforms that can have a bearing on the national economy and the way it is managed. It can affect tariffs, management, importation laws and can also lead to restructuring. This is the case in many developing countries that badly need to finance their power sector and have to borrow extensively from lending development institutions (like the World Bank) for this purpose.

It is not intended to detail each facet of project evaluation and appraisal, since these are detailed elsewhere [5–8]. The emphasis will rather be on the financial and economical evaluation. However, here are some of the aspects and activities that are investigated in the feasibility study and appraisals.

1   The sector

A study of the power sector involves a study of its development, organisation and its linkages to the rest of the economy, the institutions working in the sector, and regulatory setup. In addition, the legislation, regulations, and incentive structure inside the sector (that are likely to affect the project) are involved. The tariffs, their structure and their prospect of change and the regulatory system for setting them are included in the study, as are the sector policies and strategies, their effect and the interaction of the project with these. Taxation and importation tariffs and policies are examined for their impact on the project. During appraisals there may appear certain shortcomings in the sector or in the regulatory system that warrant pointing out.

2   *Demand (the market)*

This will deal with power demand, its past development, present growth, and future demand forecasts, and how the project will enable satisfying these. Project evaluation will look beyond the project into the system and how the electrical power system, as a whole, will interact with the project and with the availability of the new supplies and network. The demand study will also look into the shape of the demand, its timing, how it fits the load curve, and prospects for its modification. It will also look into the tariffs that apply, and if they are satisfactory; and whether the project will affect these. In the case of system strengthening projects, the study will involve an assessment of the existing detrimental financial and economic effects of the supply interruption and its consequences. Future demand and market evaluation involve predictions and, correspondingly, a measure of risk assessment that is important in the financial evaluation and in choosing the right technologies and sizes that minimise the financial risk.

3   *Technical and engineering analyses*

These are covered by the detailed studies referred to in the pre-feasibility stage. The feasibility study highlights the least-cost solution to satisfy the project objectives (usually satisfying the demand). Appraisal ascertains that the proposed technical solutions are truly the least-cost ones and that costs, timing, and implementation schedules are satisfactory. The engineering analysis looks into the project timing and implementation schedule. The technical and engineering analyses are related to the financial and economic analysis, since these are intended to evaluate the technical and engineering alternative that satisfies the demand at the least cost.

4   *Management and manpower*

This refers to the availability of technical and managerial staff to man the facilities and ensure their proper operation, maintenance, and management of supporting facilities: stores and inventories, transport, provision of services, etc. It also defines their availability and costs, the training requirements for the staff and the implication of all this on project costs.

2.2.2   *The investment phase*

Once the project is fully defined, successfully evaluated, appraised and the finance is available, the next phase of project implementation – the investment phase – begins. This has many stages.

- Carrying out the organisational, legal and financial measures to implement the project.
- Basic, as well as detailed, engineering work.
- Land acquisition.
- Tendering, evaluation of bids and contracting.
- Construction work and installation.

- Recruitment and training of personnel.
- Plant commissioning and start-up.

Good project planning and management must ensure the proper implementation of all the above stages well before the project start-up. Delays or gaps in implementation or management can cause increased costs or other damage to the utility, the investors and consumers. Execution schedules of different sections of the project need to be closely prepared, coordinated, and monitored. Typically, a network plan with identification of the critical path needs to be drawn out for the procurement, implementation, testing and commissioning of large projects. Various methods have been developed for the effective and balanced organisation of the investment phase, such as the critical path method (CPM) and the project evaluation and review technique (PERT).

This phase involves disbursement and investment expenditures which need careful assessment and evaluation. Such expenditures occur in the earliest years of project evaluation and are not significantly reduced by discounting. Therefore they have considerable effect on the project's financial viability. Their effect can be more important than future financial flows (which occur at the later stages of project operation) whose importance is greatly diluted by discounting. Such disbursements and investment expenditures have to match the financing plan of the investment phase.

2.2.3   The operational phase

If the project is well planned and executed in the pre-investment and investment phases, respectively, a few problems in the operational phase will be encountered, other than the teething problems that are not uncommon in most new facilities. The success of the project and its benefits (profitability), of course, depend not only on good engineering and management, but also on sound financial and economic evaluation, during the pre-feasibility and appraisal stages. This sound financial and economic evaluation is the subject of the next few chapters.

Electrical power facilities have a long useful operational life. To ensure that such facilities will survive their useful life demands proper operation and efficient maintenance that can involve high expenditure. All this will have been considered during the evaluation stage. However, as mentioned above, the effect of expenditures in later years of the project operation can have limited significance in evaluation owing to discounting, particularly if high discount rates are applied. It is, however, necessary to define the expected useful life of the project at the evaluation stage. This can be greatly influenced by prospects of technological change and obsolescence, as well as by changes in the fuel market and environmental legislation in the case of power generation. Such unexpected outcomes can be taken care of during risk analysis (see Chapters 13 and 14).

2.2.4   Post-operation evaluation

It is useful, at a later stage, after project completion, to revisit the project to compare the performance and results with project estimates. This mainly applies to demand, cost, execution time and evaluation of impacts, as well as returns. Such post-operation evaluation is routinely carried out by international development agencies, such as the World Bank. However, it also needs to be carried out by utilities and investors. A post-evaluation will educate the project planners and decision makers and widen their scope for future project preparation, to minimise pitfalls and risks in the preparation of other similar future projects. Unfortunately, other than for development agencies, not much post-operation evaluation is carried out. Post-operation evaluation is vital to enhance experience and reduce future risks.

2.3   Other considerations in project evaluation

Projects in the electricity supply industry, compared with other industrial and utility projects, have a few distinct features.

- They are highly capital intensive, more so than projects in any other engineering sector. They also can have long 'lead time' before they are operational. Therefore, they demand thorough planning, timing and intensive financial and economic evaluation.
- Power projects, particularly electricity generation, can have serious environmental impacts, whose mitigation can significantly affect capital cost. A thorough environmental assessment and costing are a necessity for all electrical power projects, particularly generation.
- Power projects have long useful lives; however, some of these projects may not remain operational to the end of their useful life because of technological change, public apprehension (nuclear power industry is an example), environmental regulations, shortages in fuel availability, etc., as already mentioned.
- Most of the power projects are an extension or strengthening of an existing large electrical power system and network. New large investments, like a large power station, will have a 'system effect' whose operational and financial impact extends beyond the confines of the project to affect the whole system cost. This is a consideration that has to be accounted for in large projects and demands sophisticated system analysis and simulation.

Electrical power technologies are well known and proven. Demand and markets are available and predictable, and tariffs can be regulated. Therefore the extent of risk for investments in the electricity supply industry, although significant, is lower than the average market equity. Many electricity utilities are still monopolies, although the picture is now changing in many countries. Because of their monopolistic and regulated status they are shielded from most of the market risks (and also market incentives). In such cases the risk of wrong, or non-optimal, decisions are shifted from the utility to others, such as the consumer, through tariffs. It is therefore essential that proper financial and economic evaluation, in accordance with market criteria, is carried out in order to ensure that the utility projects are viable in competition for resources with the other sectors of the economy.

The above points, as well as other considerations, affect the financial and economic evaluation of projects in the electricity supply industry. They have to be considered and will be detailed in later chapters. The introduction of computer-based design packages (on a large scale) into the design and drafting of specifications and drawings, particularly in case of power stations, have greatly reduced the lead time necessary for designing and executing a power station. Thus, significantly assisting in reducing risks and lead times for new power stations.

Also, the shorter lead times for the increasingly popular CCGT power stations and their smaller sizes, which better fit the load curve, have assisted in reducing risks of investment in new facilities. This is assisted by the increasing use of modular designs in power station projects where lead times are also considerably reduced.

The increasing role of IPPs has introduced new players into the electricity supply industry. In utilities, the risk of investment is spread over many projects. The IPP usually deals with a project at a time; therefore, risk management is paramount. This can be achieved through all parties concerned agreeing to the mutual share of financial liabilities through a security package, which involves the many agreements defining the project and outlining the obligations of the parties involved (government guarantees, power purchase agreement, fuel supply agreement, insurance, etc.). Legal arrangements are quite important in formulating such a security package. A typical project structure is shown in Figure 2.1. This diagram explains the complexity of a modern generation project [9].

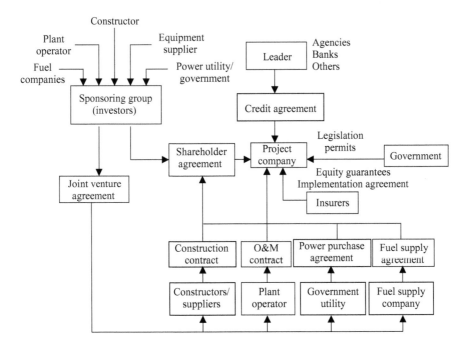

Figure 2.1   Typical project structure [9]

2.4 References

1 BREHRENS, W., and HAWRANEK, P. M.: 'Manual for the preparation of Industrial feasibility studies'. United Nations Industrial Development Organisation (UNIDO), Vienna, 1991

2 AL-BAZZAZ, M.: 'Notes on project analysis and management for planners and trainers'. Economic Development Institute – The World Bank, March 1995

3 DUVIGNEAU, J. C., and PRASAD, R. N.: 'Guidelines for calculating financial and economic rates of return for DFC projects'. World Bank Technical Paper No. 33, 1984

4 FROHLISH, E. A.: 'Manual for small industrial businesses'. UNIDO, Vienna, 1994

5 LITTLE, I. M. D., and MIRRLEES, J. A.: 'Project appraisal and planning for the developing countries' (Heinemann Education Books, 1974)

6 SQUIRE, LYN, and VAN DER TAK, HERMAN: 'Economic analysis of projects' (The Johns Hopkins University Press, 7th Printing, 1989)

7 RAY ANANDARUP: 'Cost-benefit analysis' (The Johns Hopkins University Press, 3rd Printing, 1990)

8 LAL, DEEPAK: 'Methods of project analysis' (International Bank for Reconstruction and Development, Washington, DC, 1976)

9 WHEELER, TONY.: 'Risk Management for Independent Power'. Asian Electricity Power Generation Conference, Bangkok, Thailand, 1995

Chapter 3

Time value of money (discounting)

Projects in the electricity supply industry live for a long time. As already mentioned, 25–30 years is a normal useful life for a conventional power station. The network lives even longer. For power generation projects most expenditures, in the form of operational cost (fuel, etc.) and income occur after commissioning. Such future financial flows will occur during different times and circumstances. Correspondingly, these will have different value of money than flows occurring during project evaluation. Therefore, the time value of money (discounting) and the choice of a proper discount rate is highly important for capital intensive long-life projects with large operational cost, like those of the electricity supply industry.

3.1 Discounting

Generally, *financial flows* (streams of expenditures and income) from projects do not occur during the project evaluation [1,2]. They occur after a year, a few years, or often after many years. Annual *cash flows* are the difference between money received and money paid out, each year. Cash flows or financial flows (outlays) occurring at different times cannot be readily added, since £1 today is different from £1 next year, and very different from £1 in 20 years' time. Before dealing with financial flows occurring at different times, these have to be adjusted to the value of the money at a specified date, which is normally called the *base date* or *base year*.

Therefore, an important factor to recognise in project evaluation is the time value of money. A pound today is more valuable than £1 tomorrow and £1 tomorrow is more important than £1 the day after, etc. There are many reasons for this.

- Future incomes are eroded by inflation; therefore the purchasing power of a pound today is higher than a pound in a year's time.
- The existence of risk, an income or expenditure that occurs today, is a sure amount. Future income or expenditure may vary from anticipated values.
- The need for a return; by undertaking investment and foregoing expenditure today, an investor expects to be rewarded by a return in the future.

An entrepreneur expects to gain a premium on his investment, to allow for the following three factors: inflation, risk taking, and the expectation of a real return. That is, he expects to regain his money, plus a return that tallies with the market and his estimation of these three factors.

Even if inflation is allowed for, or ignored, money today will still remain more valuable than tomorrow's money because of risk and expectation of a reward by forgoing today's expenditure. To an investor, £1 today is more valuable than tomorrow's £1, because it can be invested immediately and can earn a real income, that is, a return higher than inflation. Today's £1 will equal tomorrow's £1 plus a real value.

Present valuing (discounting) is central to the financial and economical evaluation process. Since most of the project costs, as well as benefits, occur in the future, it is essential that these should be discounted to their present value (worth) to enable proper evaluation. Present valuing will be carried out through discounting next period's financial outlay (F_1) to its present value through multiplying it by a *discount factor*. The discount factor is a function of the *discount rate* (r), which is the reward that investors demand for accepting a delayed payment. It is also referred to as the *rate of return* or *opportunity cost of capital*, so that

$$\text{present value (PV)} = \text{discount factor} \times F_1$$

where discount factor $= 1/(1+r)$. The discount factor is sometimes called the *present worth factor*. It is defined as how much a pound in the future is worth today. Therefore, with a discount rate (expected rate of return) of 10 per cent annually, the discount factor for the first year's financial outlay will be $1/(1 + 0.1) = 0.909$, and £110 materialising after one year will equal to $0.909 \times £110 = £100$ today.

Similarly, an outlay at year 2 will have to be multiplied by $1/(1 + r)^2$ and that occurring at a year n will have a discount factor of $1/(1+r)^n$. Therefore, the discount factor, in year n, is equal to

$$\text{discount factor} = 1/(1 + r)^n$$

and present value $= F_n \times$ discount factor $= F_n \times [1/(1 + r)^n]$. Therefore, £1000 occurring after 5 years, with a discount rate of 10 per cent, will have a present value equal to £1000 $\times [1/(1 + 0.1)^5] = £620.92$ today. Similarly, £1000 occurring after 30 years will be equal to $£1000.0 \times [1/(1 + 0.1)^{30}] = £57.31$ today.

In the same way, the process of discounting can be converted to a process of compounding when it is required to present value past payments. The *compound factor* is the reciprocal of the discount factor and is equal to $(1 + r)^n$. Therefore, a financial outlay that occurred one year earlier has a present worth equal to $(1 + r)$ of the value of that outlay today. A payment of £1000 that occurred a year from now, will have a present value of $£1000(1 + 0.1) = £1100$ today.

Therefore, the discount factor $[1/(1 + r)^n]$ is universal, with n positive $(+n)$ for all future years and negative $(-n)$ for all past years' financial outlays, with $n = 0$ for the base year. For the ease of the analysis the term *discounting* will also mean *compounding*, taking into account the timing of the outlay as mentioned above.

It has to be emphasised that present valuing is based on *compounding returns*. That is, the expected return (say interest rate) on the first year's money is also added to the original investment and then the sum reinvested at the same interest rate. This is different from *simple return* (simple interest), where interest is only paid on the original investment. Therefore, in simple interest, with an interest rate of 10 per cent, a £100 invested today is equal to £100 + £10 after one year, £100 + £10 + £10 after two years, and £130 after three years, etc. In compound interest payments, the £100 investment will equal to £100(1 + 0.1)², £100(1 + 0.1)³, that is, £121 and £133.10 after two and three years, respectively. Compounding and not simple returns are the basis for present valuing.

When the discounting or compounding procedure is applied to money, the rate (r) is often referred to as a *rate of interest*. When the procedure is applied to economic resources in a more general sense (the electricity supply industry), it is usually referred to as the *discount rate* [3]. The discount rate is, therefore, a more general term than the interest rate. In most of the chapters of this book the term discount rate will be used.

3.2 Discounted cash flows

It is better if the net present valuing is stated in terms of discounted cash flows. A cash flow has already been defined as the difference between money received and money paid. Each year's future cash flows can be discounted to their present value by dividing them by the discount factor for that year.

Therefore, extended stream of cash flows $M_0, M_1, M_2, \ldots, M_n$ occurring at years $0, 1, 2, \ldots, n$ has a present value of:

$$PV = M_0 + \frac{M_1}{(1+r)} + \frac{M_2}{(1+r)^2} + \cdots + \frac{1}{(1+r)^n} = \sum_N \frac{M_n}{(1+r)^n}.$$

In the special case of $M_1 = M_2 = \cdots = M_n = M$

$$PV = M \sum_N \frac{1}{(1+r)^n}.$$

To give a simple example, consider a short-term project involving an investment of £50 000 thousand at the beginning of each year over four years, starting today, with a discount factor of 8 per cent. Its present value is equal to

$$PV = 50 \times 10^3 \left[1 + \frac{1}{1+0.08} + \frac{1}{(1+0.08)^2} + \frac{1}{(1+0.08)^3} \right]$$

$$= 50 \times 10^3 (1 + 0.926 + 0.857 + 0.794) = £178\,850.$$

As already mentioned, present valuing is a process of compounding returns. Therefore, the same principles of discounting (compounding) are applicable for past payments, and their present valuation. For instance, a cash flow 'M' that occurred in a past year 'n', has a present value of $M_n(1+r)^n$ in today's money. The present value of £100 paid three years before is equal to $£100(1+0.1)^3 = £133.1$ at a discount (compound) rate of 10 per cent annually.

In the above example, if the investments occurred in the past four years, their present value today will equal:

$$PV = £50 \times 10^3[(1+0.08)^4 + (1+0.08)^3 + (1+0.08)^2$$
$$+ (1+0.08)] = £243\,300.$$

This approach is quite useful in electrical power projects since, in some project evaluations, the base year (i.e. the year 0, to which all payments are discounted) is usually the year of commissioning. Therefore, future cash flows are discounted to that year, while past flows are also compounded (discounted) to that commissioning year.

Because of discounting, the value of future cash flows becomes eroded significantly year after year, particularly if discount rates are high. In the above example, at a discount rate of 10 per cent, a payment occurring in 30 years will have a present worth of only 5.7 per cent of its value. If the discount rate is 15 per cent, then this present worth is only 1.5 per cent, which is almost negligible. Therefore, in project evaluation, the financial streams of the first few years have great significance. These need to be predicted with accuracy, because it is the money streams of these early years that have greater real value and impact. Discounting greatly helps in reducing the significance of inaccuracies in predicting future long-term cash flows. However, this entirely depends on the discount rate. A low discount rate will help to retain a significant value of future cash flows, while a high discount rate will render long-term cash flows almost valueless today.

Discount tables, amortisation ratios, etc., are readily available in the literature [3,4], and through modern financial calculating machines.

3.3   Net present value (NPV)

Along the life of the project there will be two financial streams: one is the costs stream (which includes capital and operational cost (C)) and the other is the benefits stream (B). The two streams must contain all costs and benefits for the same estimated life frame of the project. The costs stream, being outward-flowing cash, is regarded as negative. The difference between the two streams is the cash flows – the '*net benefits stream*'. The values of the net benefits in certain years can be negative, particularly during construction and the early years of the project.

Discounting the net benefits stream into its present value, by multiplying each year's net benefits by that year's discount factor, will present the net present value (NPV) of the project [5,6], as detailed in the example in Table 3.1, which utilises a discount rate of 10 per cent. Notice that outward-flowing cash (costs) are negative

Table 3.1   NPV – real terms

I Year	II Cost	III Income	IV = III – II Net benefits	V Discount factor	VI = V × IV Net benefits discounted at 10%
−1	40	—	−40	1.100	−44.00
0	110	40	−70	1.000	−70.00
1	10	40	30	0.909	27.27
2	10	40	30	0.826	24.79
3	10	40	30	0.751	22.54
4	—	70 (salvage value)	70	0.683	47.81
				Net present value: £8.41	

whereas inward-flowing cash (income) is positive:

$$\text{NPV} = \sum_N [B_n - C_n]/(1 + r)^n.$$

Usually projects are undertaken because they have a positive net present value. That is, their rate of return is higher than the discount rate, which is the opportunity cost of capital. The calculation of net present value is the most important aspect in project evaluation and its positive estimation, at the designated discount rate, is essential before undertaking a project.

3.4   Accounting for inflation in cash flows

Future cash flows can be estimated either in today's money and are usually termed *cash flows in real terms*, or in the money of the date (year) in which they occur and are termed as *monetary or nominal cash flows*. It is usually easier to predict future cash flows in real terms, i.e. today's money, because today's information is the best available. Utilising real terms will eliminate the need to predict another future factor – inflation – thereby reducing the number of the unknowns.

Nominal cash flows are quite important for commercial financial statements and accounts. They are used in the prediction of financial statements allocations (depreciation, interest, etc.), in order to arrive at gross profits and net profits (after allowing for taxation and similar outlays). To arrive at nominal cash flows, predicted real cash flows have to be inflated by the expected inflation rate.

In present valuing, a real discount rate has to be utilised if cash flows are in real terms (today's money). A nominal (monetary) discount rate is utilised if cash flows are in monetary terms that are in the money value of the year in which they occur. A nominal discount rate is equal to the real discount rate modified by the inflation rate:

$$\text{nominal discount rate} = [(1 + r)(1 + \text{inflation rate})] - 1.$$

Table 3.2   NPV – nominal terms

Year	Nominal cost	Nominal income	Nominal net benefits	Nominal discount factor	Net benefits discounted at 15.5%
−1	38.1	—	−38.1	1.155	−44.01
0	110.0	40.0	−70.0	1.000	−70.00
1	10.5	42.0	31.5	0.866	27.27
2	11.0	44.1	33.0	0.750	24.79
3	11.6	46.3	34.7	0.649	22.54
4	—	85.1	85.1	0.562	47.81
				Net present value: £8.41	

Suppose, for instance, that a real discount rate, which allows for real return and a risk of 10 per cent, is adopted. Inflation is expected to be 5 per cent in the future.

Then the nominal discount rate will equal

$$[(1 + 0.1)(1 + 0.05)] - 1 = 0.155, \text{ i.e. } 15.5 \text{ per cent.}$$

Because of their small values and as an approximation, a nominal discount rate equals the real discount rate plus inflation. If cash flows are in real terms then real discount rate is utilised. If they are in nominal terms then the nominal discount rate is employed. Net present value is the same when utilising both methods.

The cash flows of Table 3.1 can be presented in nominal terms by inflating them by the annual inflation rate of 5 per cent as in Table 3.2.

This is the same result as that of Table 3.1, which indicates that the real and nominal cash flows will give the same net present value if the real and nominal discount rates are properly utilised with each cash flow respectively.

It has to be noted that the expenditure in the year (−1) in Table 3.1 was £40 in the money of the base year (year 0). This is only equal to £38.1, i.e. [£40 ÷ (1 + 0.05) inflation rate], in the money of year −1. The same applies to the estimate of the salvage value, which is estimated at £70 in money of the base year and is equal to $(70 \times [1 + 0.05]^4) = £85.1$ in the money of year 4.

In most of the analysis of this book, it is the real cash flows and the real discount rate that will be used to discount streams of income and expenditure (past as well as future). When values are available in the money of the day of the transaction (like fixed rent values, payments for fixed rates and tariffs, fixed price fuel contracts, etc.), then these values have to be deflated to the base-year money, and then discounted by the real discount rate to their present worth.

In many power-generation projects it is the commissioning year of the project that is termed as the base year. Prior to the base year many payments, usually project investment cost, have already been incurred. These have first of all to be presented in the money of the commissioning year, then to be compounded (discounted) to their present value by multiplying them with the compounding factor $(1 + r)^n$. This is the same as multiplying them by the discount factor $1/(1+r)^{-n}$, where n here is negative because it is prior to the base year. Cash streams occurring after the base year (commissioning date), have to be presented in the base-year money and discounted by multiplying them with the discount factor $1/(1+r)^n$. Therefore, the discount factor $[1/(1+r)^n]$ is universal for all cash flows with n as negative for all flows prior to the base year, positive for all flows after the base year, and zero for the base year. For the ease of treatment, the term 'discounting' will be universally used for both 'compounding' and 'discounting'.

In the past analysis, we have assumed that price changes of operating costs as well as revenues are going to change over time (inflate) at the same rate. This may not be true in many cases, since relative price changes do often occur. Different cost categories, like fuel costs, may have different rates of change over time from other costs like labour or other materials. In this case, future streams are presented in nominal terms utilising each cost or income item, and its expected inflation rate. Real cash flows are obtained by deflating these by the average annual inflation rate expected during the projected period. Alternatively, the project cash flows are presented in real terms with the stream of the cost or income item(s) expected to significantly deviate from the average annual inflation rate, inflated (deflated) by the inflation differential between its anticipated inflation rate and that of the average annual inflation rate.

3.5 Considerations in present valuing

The above analysis indicated that projects consist of financial streams of benefits (B) and costs (C), which occur at different years through the life of the project. Therefore, it is essential to understand how to deal with these streams, manipulate them to allow for the computation of their net present value, and allow for comparison of the different benefits and costs. This section details various present valuing (discounting) formulas and their utilisation. It recapitulates on the above analysis and introduces new concepts. Future and past cash flow are termed M, and the discount rate is termed r. These are all in real money, that is ignoring inflation (if they are represented in nominal terms then the discount rate must incorporate the inflation rate as detailed above). The following are general rules that need to be understood by all project evaluators, although all do not necessarily apply to the electricity supply industry.

3.5.1 Future and past valuing

Future valuing (FV) of a present value (PV) means that the base year has been moved into the future by n years. Therefore, the PV is occurring now at $(-n)$ years, from the new base year. The universal discount (compound) factor is maintained, with negative n value, i.e. $FV = PV \times [1/(1+r)^{-n}] = PV(1+r)^n$. A similar approach applies for past valuing, which means that the base year has been moved into the past by n years. The past value will equal $PV \times [1/(1+r)^n]$.

3.5.2   Annuity factor

Present valuing of a stream of equal cash flows M:

$$PV = \frac{M}{1+r} + \frac{M}{(1+r)^2} + \cdots + \frac{M}{(1+r)^n} = M \sum \frac{1}{(1+r)^n}.$$

If we substitute 'a' for $M/(1+r)$ and 'x' for $1/(1+r)$ then

$$PV = a(1 + x + x^2 + \cdots + x^{n-1}).$$

Multiplying both sides by x we have:

$$xPV = a(x + x^2 + \cdots + x^n).$$

Subtracting the second equation from the first:

$$PV(1 - x) = a(1 - x^n).$$

Substituting for a and x and then multiplying both sides by $(1+r)$ and rearranging gives:

$$PV = M \left[\frac{1}{r} - \frac{1}{r(1+r)^n} \right].$$

The expression in square brackets in the above equation is the *annuity factor*, which is the present value of an annuity £1 paid at the end of each of n periods, at a discount rate r:

$$\text{annuity factor} = \left[\frac{1}{r} - \frac{1}{r(1+r)^n} \right].$$

Therefore, the annuity factor is the summation of all the annual values of the discount factors over the period, i.e. annuity factor $= \sum_N$ discount factor, and

$$PV = M \times \text{annuity factor}.$$

Thus, the present value of annual payments of £20 each, paid for 10 years, with the first occurring after one year, and at a discount rate of 8 per cent, is equal to

$$PV = £20 \left[\frac{1}{0.08} - \frac{1}{0.08(1 + 0.08)^{10}} \right] = £20 \times 6.710 = £134.20.$$

In the above equation of the annuity factor, if the cash flow is maintained to perpetuity, then the annuity factor is reduced to $1/r$ and the present value of the *perpetuity*:

$$PV = \frac{M}{r}.$$

In the above example, if a bond has an annual payment of £20, which is permanently maintained, then the present value of all future payments will be

$$PV = £20/0.08 = £250.$$

If the above annual payments to perpetuity are increased year after another, by an annual percentage '*g*' (like reinvesting the annual return by *g* per cent) then the present value of the perpetuity becomes:

$$PV = \frac{M}{1+r} + \frac{M(1+g)}{(1+r)^2} + \frac{M(1+g)^2}{(1+r)^3} + \cdots .$$

It can be proven that the above equation, over a long period, can be reduced to:

$$PV = \frac{M}{r - g}$$

Therefore, and with the same assumptions of the above example, if we consider a bond that pays 4 per cent in annual real return, and if this return is reinvested then the PV of the perpetuity becomes:

$$PV = £20/(0.08 - 0.04) = £500.$$

The annual rate of return of *r*, compounded *m* times a year, amounts by the end of the year to $[1 + (r/m)]^m - 1$.

Thus, an annual return of 12 per cent, paid as 1 per cent monthly, and with a return at the same rate compounded, is equivalent to $[1 + 0.12/12]^{12} - 1 = 12.68$ per cent annually.

Generally speaking, if the benefits are spread continuously, then the above equation of $[1 + (r/m)]^m$ approaches e^r as *m* approaches infinity, where e is the base for natural algorithms and is equal to 2.718. For a continuously compounded benefit or cost the end of the year value has to be multiplied by $(2.718)^r$. Therefore, if the interest in the above example is paid daily and reinvested, the compounded return will equal approximately 12.75 per cent annually.

3.5.3   Capital recovery factor (CRF) or equivalent annual cost

An annuity factor is a means of converting a stream of equal annual values into a present value, at a given discount rate (interest). *A capital recovery factor (CRF)* performs the reverse calculation. It converts a present value into a stream of equal annual payments over a specified time, at a specified discount rate (interest).

From the above equation of the annuity factor, it is possible to derive the CRF or the *equivalent annual cost*. This is the amount of money to be paid at the end of each year 'annuity' to recover (amortise) the investment at a rate of discount *r* over *n* years.

The equivalent annual cost *M* will be the reciprocal of equation of PV mentioned earlier, i.e.

$$M = \frac{PV}{\text{annuity factor}} = PV \times CRF$$

and

$$CRF = \frac{1}{\text{annuity factor}} = 1 \Big/ \left[\frac{1}{r} - \frac{1}{r(1+r)^n} \right] .$$

For example, the equivalent annual capital cost of an investment of £1 million over 10 years, at a rate of interest of 12 per cent will be:

$$= £1\,000\,000 \left/ \left[\frac{1}{0.12} - \frac{1}{0.12(1+0.12)^{10}} \right] \right.$$

$$= £1\,000\,000 \times 0.176\,98 = £176\,980 \text{ annually.}$$

This is the same as the annual mortgage payment to acquire a house. Thus a house costing £1 million, which has to be repaid over 10 years with an interest (mortgage) rate of 12 per cent, will entitle 10 annual payments of £176 980 each.

To prove that the CRF is the reverse of the annuity factor, consider these equal annual payments of £176 980 discounted over 10 years, at a rate of interest of 12 per cent. Their present value will be:

£176 980 × annuity factor.

The annuity factor from the Present Value Tables is 5.6502, so

£176 980 × 5.6502 = £1 000 000,

which is the original investment.

When such payments are executed on a monthly basis, with a monthly rate of interest equal to 0.01 per cent, and the payments are spread over 120 months, the above equation becomes:

$$£1\,000\,000 \left/ \left[\frac{1}{0.01} - \frac{1}{0.01(1+0.01)^{120}} \right] = £14\,347 \text{ monthly.} \right.$$

For semi-annual payments the annual fee will be:

$$£1\,000\,000 \left/ \left[\frac{1}{0.06} - \frac{1}{0.06(1+0.06)^{20}} \right] = £87\,184 \text{ semi-annually.} \right.$$

The CRF is the sum of two payments. The first is repayment of the principal (amortisation) and the second is the interest on the unrepaid principal.

The annual CRF for £1 at an interest rate of 12 per cent and a period of four years is 0.3292, as shown in the following schedule.

	Principal	Interest	CRF (total)
1	0.2092	0.1200	0.3292
2	0.2343	0.0949	0.3292
3	0.2624	0.0668	0.3292
4	0.2941	0.0351	0.3292
	1.0000	0.3168	

It is very important to understand both the CRF and the annuity factor. They are useful in daily financial life and in quick comparison of projects and evaluation of alternatives in the electricity supply industry, as will be demonstrated in Chapter 5.

3.5.4 Grouping monthly and hourly flows

Not all costs and benefits occur at the end of the year. The main benefit of electricity production, which is generation in kilowatt-hours (kWh), occurs continuously every hour of the year. To accumulate these as a single generation figure at the end of the year is an inaccurate approximation. Payments for salaries, fuel, etc., are costs that are incurred monthly and continuously throughout the year. To lump them as a single payment at the end of the year and to discount them will result in an underestimation. To overcome this, such financial flows can be presented in a monthly (or hourly) form, which makes the calculation cumbersome. As an alternative, they can be lumped as an annual flow at the middle of the year, which is a useful approximation.

As an example, consider generation from a plant of capacity of 1 kW. Its annual continuous production is 8760 kWh, spread throughout the year, with 1 kWh every hour. If this production is grouped as one figure at the beginning of the year, with an annual discount factor of 12 per cent (hourly discount rate of 12 per cent/8760), it will be equal to:

$$1 \bigg/ \left[\frac{0.12}{8760} \right] - 1 \bigg/ \left[\frac{0.12}{8760} \left(1 + \frac{0.12}{8760} \right)^{8760} \right] = 8255 \, \text{kWh}.$$

Grouping this at the end of the year will have a present value equivalent to:

$$1 \bigg/ \left[\frac{0.12}{8760} \right] - 1 \bigg/ \left[\frac{0.12}{8760} \left(1 + \frac{0.12}{8760} \right)^{-8760} \right] = 9307 \, \text{kWh}.$$

Neither of these figures tallies with the annual generation of 8760 kWh, which is recorded by the operators.

If both the figures above are discounted to mid-year at 12 per cent annual discount rate they will be 8787 and 8765 kWh, respectively. Therefore, grouping all the annual generation in its physical terms as one lump sum, in the middle of the year, is a useful approximation. The same applies to other expenses (payments for fuel, salaries, benefits, etc.), which are approximately evenly distributed throughout the year. This is the procedure, which was adopted by UNIPEDE in its comparison of the cost of different generating facilities [7].

Therefore, if the commissioning year is the base year, then the cash flow of the first year has to be grouped at the middle of the first year for present valuing purposes. The second year's cash flow is grouped at 1.5 years from the base year, etc. Deviation from this, like the common practice of grouping all cash flows at end of the year, causes errors.

3.6 References

1 GITTINGER, J. R.: 'Economic analysis of agricultural projects' (The John Hopkins University Press, 1982, 2nd edn.)

2   BREALEY, R., and MYERS, S.: 'Principles of corporate finance' (McGraw-Hill, 2000)

3   'Discounting and Measures of Project Worth', Development and Project Planning Center, 1995, University of Bradford

4   'The world measurement guide', *The Economist*, London 1980

5   KHATIB, H.: 'Financial and economic evaluation of projects', *Power Eng. J.*, February 1996

6   DUVIGUEAU, J. C., and PRASAD, N.: 'Guidelines for calculating financial and economic rates of return for DFC projects', International Bank for Reconstruction and Development, Washington DC, 1984

7   'Electricity generating costs for plants to be commissioned in 2000' (UNIPEDE, Paris, January 1994)

Chapter 4

Choice of the discount rate

4.1 Introduction

The life cycle costs of a project and its feasibility, for a given output, depend on three factors: (i) the investment cost, (ii) the operational costs, and (iii) the discount rate utilised. Many planners think that the discount rate is the most important of these three factors. It greatly affects the whole economics of the project and the decision making, particularly in capital-intensive projects like those of the electricity supply industry [1]. The discount rate almost governs the choice of the least-cost solution. It also greatly affects estimation of the net returns from the project (net present value) during the evaluation stage, the project's feasibility, and the decision to proceed with the investment or not. A high discount rate will favour low capital cost with higher operational cost project alternatives. A low discount rate will tend to weigh the decision in favour of the high capital cost and low operational cost alternatives.

In spite of its crucial importance in project evaluation, it is surprising how little effort project evaluators exert to research the proper discount rate needed for project evaluation. Simultaneously, all the effort in estimating investment and operational costs is rendered worthless by a deviation in the choice of the discount rates. Many project evaluators usually take a specific discount rate, of their own choice, as appropriate and then try to cover up for its possible inaccuracies by sensitivity analysis. Sometimes, owing to lack of clarity about discount rates, two (or more) discount rates are chosen for evaluation [2]. Often, for government-sponsored projects, the work of the planner is made easier by the authorities fixing the discount rate (sometimes inaccurately).

In many cases, the internal rate of return (IRR) of the project is calculated and if this is considered to be appropriate by the investors (utilising their experience, hindsight and possible available returns and risks of other projects), then a decision is taken to proceed with the investment without having to resort to the detailed calculation of an adequate discount rate. However, such a procedure does not allow the calculation of the net present value of a project, or adequate comparison of different alternatives (see Chapter 5).

4.2   The discount rate

The discount rate is the opportunity cost of capital (as a percentage of the value of the capital). The opportunity cost of capital is the return on investments forgone elsewhere by committing capital to the project under consideration. It is also referred to as the *marginal productivity of capital*, i.e. the rate of return that would have been obtained by the last acceptable project. In investment decisions, the opportunity cost of capital is the *cut-off rate*, below which it is not worthwhile to invest in the project [3].

The discount rate connotes the entrepreneur's indifference to the timing of the return. If it is equal to 10 per cent, then the entrepreneur is indifferent to whether he receives £1 today or £1.10 a year from today. This indifference is the basis for engineering economics. To serve this purpose, the nominal discount rate should at least be equal to a value which, after tax, would compensate the entrepreneur for the following three objectives, which have already been mentioned in Chapter 3: (i) reduction in the purchasing power of money which is brought about by inflation, (ii) provision of a real return, (iii) compensation for the extent of risk undertaken by committing capital to this investment. The value of the nominal discount rate is correspondingly a function of the above three factors: inflation, risk-free real return, and the extent of risk in the project. A real discount rate, which ignores inflation, is utilised if the cash flows are presented in the base-year money, as explained in Chapter 3. When reference is made just to the 'discount rate', then it is the real discount rate that is meant.

To demonstrate the importance of discount rates, the cases for nuclear and coal power station were compared. The results of the evaluation are demonstrated in Figure 4.1. The figure gives the price per kWh of output and shows how the economics

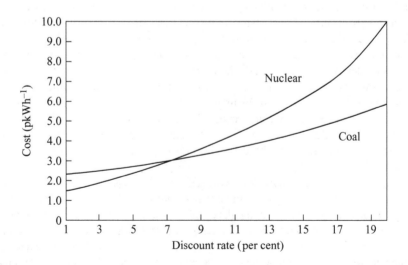

Figure 4.1   Sensitivity of nuclear and coal power stations costs to discount rate [4]

of each alternative changes with the discount rate. At low discount rates less than 7 per cent, the nuclear alternative is cheaper. However, at a discount rate of 9–10 per cent or higher, it is definitely the coal alternative that is in favour, even at high coal prices. Obviously, the discount rate is crucial in such decision making between alternatives [4].

	Nuclear	Coal
Investment cost (£ million)	1527	895
Operating costs kWh^{-1}	0.37p	0.46p
Fuel cost kWh^{-1}	0.45p	1.43p

From Figure 4.1 it is clear how, for evaluation purposes, the cost per kWh generated is greatly affected by the discount rate for high capital-intensive investments, like those of a nuclear power station. The cost of 2p kWh^{-1} more than doubled when the discount rate was increased from 4 per cent to less than 11 per cent. Between these two discount rates, the cost of each kWh from a coal power station, which is less capital intensive but has a much higher operational cost, increased by less than one third.

Figure 4.1 demonstrates the important fact of the sensitivity of different types of investment to the choice of the discount rate. It is clear that, because future net benefits are greatly reduced by the higher discount rate, the cost per kWh rapidly rises for capital-intensive nuclear power stations more than it does for the less capital but higher operating cost coal power stations. For low operating cost alternatives, like nuclear, the high net benefits are severely eroded by the high discount rate, while the high front capital investment is compounded by the high discount rate to the base year of commissioning. For the higher operating cost alternative, the net benefits as well as the front investment are smaller, and correspondingly the result is less affected by discounting. The choice of the proper interest rate is crucial in such highly different investment cost alternatives.

Because of the primary importance of choosing the right discount rate in any project evaluation, this subject is dealt with in some detail in this chapter. In this analysis, which is carried throughout utilising the project's real costs and benefits, at the base year prices, we are only concerned with the real discount rate (i.e. ignoring inflation). At this stage, taxation is also ignored.

4.3   Calculating the discount rate

In most countries, projects financed by the government use a different discount rate than those used by the private sector investors. Normally, government investments are less risky, because they are mostly in regulated utilities and industries. The discount rate of the private sector investments is influenced not only by risk, but also by returns in the bond market which can change significantly from one period to another. Both discount rates are, however, significantly influenced by availability of capital for investment and the cost of borrowing.

4.3.1   Government financed projects

For public projects, the discount rate is sometimes decided by the responsible institutions in the government, which greatly eases the work of the planner of government sponsored projects. In the 1978 White Paper guidelines for nationalised industries, the UK Treasury indicated that the required rate of return for the nationalised industries projects should be 5 per cent in real terms (this was at a time in which the real returns of risk-free investment, like those of short-term government bonds, was very close to zero). Later, in April 1989, this was increased to 8 per cent and in the mid-1990s this required rate of return was adjusted to 6 per cent annually [5]. In other countries, hindsight is employed. The cut-off rate is in the background thinking of the decision maker. If the rate of return falls below a certain level (say, 10 per cent), the project is deemed unacceptable. If it is higher than another level (15 per cent, for instance), the project is held in great favour. The level of such rates depends on the availability of capital and public funds for the government utilities to invest, and the number and value of competing projects. It also depends on the availability of loans for the public utilities and the rate of interest of such borrowing. It is usually appropriate for the government to undertake the project as long as its rate of return is higher than the real cost of borrowing. The actual cost of borrowing takes into account the budget deficit, the indebtedness of the government; that is, the public local debt as well as foreign, and the capability of the government to repay this in the future as well as other macro-economic considerations relevant to the role of the government in the economy. Sometimes this discount rate is referred to as the social discount rate.

International development agencies, such as the World Bank and other regional development banks, also consider, during project appraisal, the macro-economic situation of the borrowing developing country to assess the right discount rate to apply in the case of that particular country. In their studies of the economics of nuclear electricity generation, UNIPEDE/EURELECTRIC chose what it considered to be two logical real discount rates, namely 5 and 10 per cent, and evaluated the economies of different generation facilities utilising both of those discount rates. Of course, this is only indicative. For decision making, a single discount rate has to be chosen as a reference discount rate. This discount rate has to take into consideration most of the points explained below. In case of business decisions it must be equal to the opportunity cost of capital.

4.3.2   Business investment projects

Investors expect a rate of return from projects to compensate them for the following: a minimum acceptable real return available in the market (risk-free rate of interest), the risk of investing in the project, taxation and also inflation. As explained above, the rate of return will be calculated in real terms thus ignoring inflation. Therefore, the real returns from the project are estimated to compensate the entrepreneur for the following factors.

Table 4.1   *Average rates of return on treasury bills, government bonds, corporate bonds, and common stocks, 1926–1999 (figures in annual percentages)*

Portfolio	Average annual rate of return		Average risk premium (extra return versus treasury bills)
	Nominal	Real	
Treasury bills	3.8	0.7	0
Government bonds	5.6	2.6	1.8
Corporate bonds	6.1	3.0	2.3
Common stocks (S&P 500)	13.0	9.7	9.2
Small-firm common stocks	17.7	14.2	13.9

Source: Ibbotson Associates, Inc., Yearbook.

(i) The risk-free rate of interest

The risk-free rate of interest measures the time value of money. The best measure for measuring market risk-free rates of interest is the index-linked gilt. These government index-linked bonds (linked to the cost-of-living index) will yield to their owners, annually, an interest that is equal to a risk-free rate of return and also compensate them for inflation. It is also possible to calculate the real risk-free rate of interest by deducting inflation from the nominal yield on conventional government bonds.

Table 4.1 features the average return on treasury bills and government bonds in the United States, as well as US common stocks until 1999. From this table it is clear that government bonds have a real return that averages around 2.6–3 per cent annually, while the common stock has a risk premium of around 7 per cent above government bonds.

A return of around 3 per cent is therefore the minimum real return expected by an investor when undertaking a risk-free investment, like putting money into government bonds.

(ii) Premium to compensate for risk

Equity is an ownership right or risk interest in an enterprise. Return from equity (stocks) involves capital gains (or losses) as well as dividends. Investing in equities entails risk, equity prices, as well as dividend fluctuations. Therefore, investors in the financial market (stock exchange) expect a premium over investors in government bonds to compensate them for the risk they are undertaking. The amount of this premium depends on their and the market's evaluation of the risk of the investment, as well as the availability of other remunerative outlets for their money.

Return on equities (capital gains plus dividends) fluctuates in the stock market. However, over the long term, the average equity expects to have a higher real return than bonds. Otherwise, there will be no incentive to invest in the risky stock exchange. The arithmetic mean of the return on the value-weighted index of all equities (the market portfolio) in the UK during the period 1955–1994 was 10.9 per cent on the average annually, but this has dropped in recent years.

An almost similar picture exists in the USA, where the average risk premium (extra real return of stocks over treasury bills) was around 9 per cent annually over the lengthy period of 1926–2000 [6]. This figure similarly tallies with that of the UK capital market experience, as well as other markets. Therefore, an average risk premium of 8–9 per cent (say 8.5 per cent) above treasury bills, or 7 per cent above government bonds is considered reasonable for the average equity.

The above situation applies for the average equity, i.e. to the market portfolio. However, each category of investment has its own risk measure; investment in new business ventures, depending on its technologies and markets, is much more risky than the average equity. Simultaneously, regulated utilities have a risk, which is lower than the average market risk. A stock's sensitivity to change in the value of the market portfolio is known as *beta*. Beta, therefore, measures the marginal contribution of a stock to the risk of a market portfolio. In a competitive market, the expected risk premium varies in direct proportion to beta. This is the capital asset pricing model (CAPM) [6], simply defined as

expected risk premium on a stock = beta × expected risk premium on market.

Therefore, treasury bills have a beta of zero because they do not carry out any risk. The average equity in the market portfolio has a beta of 1. Therefore, its risk premium is the average market risk premium of 8.5 per cent.

Generally speaking:

expected risk premium on investment = beta × average market risk premium

and real discount rate = real risk-free rate + (market risk premium × beta).

Therefore, investment in an asset that has a beta of 0.6 means that the real discount rate for this investment will be equal to 5.8 per cent (0.7 per cent (which is the risk-free rate) + (8.5 per cent market risk premium × 0.6)).

It is therefore necessary to discuss the beta of the electricity utilities. Utilities, in many cases, are monopolies. They have well-defined markets and also established technologies; correspondingly they have a lower beta than the average equity. Recent beta for electricity generating companies in the UK was around 0.9 and that of the Regional Electricity Companies was around 0.8. Beta of the mean equity for electric generating utilities in the USA was much lower, at an average of 0.51. Betas for Japanese utilities fluctuated much more than that of UK and USA utilities. They were as low as that of the USA utilities until the mid-1980s, then increased, and were above 1.0 in recent years. Table 4.2 gives the beta of several large USA electric utilities. Mean equity betas of American and Japanese electric utilities are in Figure 4.2.

Apparently, no general global figure can be put into the betas of electricity utilities. Generation utilities have a higher risk and correspondingly higher beta value than distribution utilities. Nuclear installations have a higher risk than other forms of thermal generation and, correspondingly, utilities with nuclear generation have a higher

Table 4.2 Betas for some large electric utilities in the USA

Firm	Beta	Standard error
Boston Edison	0.60	0.19
Central Hudson	0.30	0.18
Consolidated Edison	0.65	0.20
DTE Energy	0.56	0.17
Eastern Utilities Associates	0.66	0.19
GPU, Inc.	0.65	0.18
New England Electric System	0.35	0.19
OGE Energy	0.39	0.15
PECO Energy	0.70	0.23
Pinnacle West Corp.	0.43	0.21
PP & L Resources	0.37	0.21
Portfolio Average	0.51	0.15

Source: The Brattle Group, Inc.

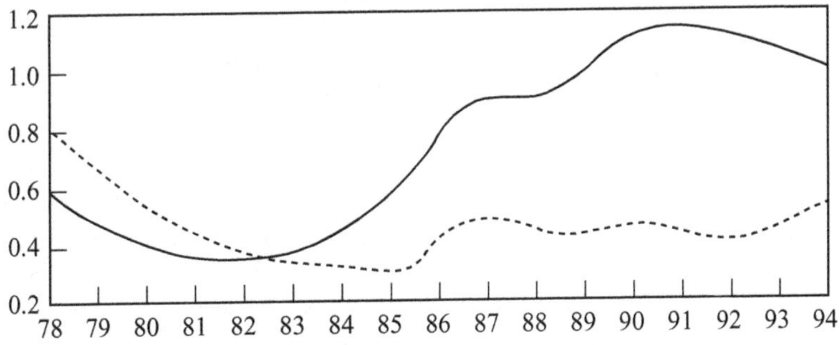

Figure 4.2 Mean betas of electric utilities in US and Japan [7]

- - - 69 US operators
— 9 Japanese operators

beta than other generating utilities, depending on the extent of their nuclear component. Long-life facilities, like large coal-firing plants, carry more risk than modern CCGT gas-firing facilities. Investments in big long-lead-time pulverised coal firing generating units is riskier than investing in smaller short-lead-time CCGT plants that easily fit the load curve. Therefore, the risk of investment in electricity utilities in a particular country compared with other equities in the financial market depends on the extent and type of regulation in the utility market in that country, the sort of business of the utility (generation or distribution or both), the type of facilities utilised and their expected life. Generally speaking, regulated utilities have a lower beta than the average equity. Betas between 0.4 and 0.9 are normal for utilities depending on the type of business and extent of regulation. The regulatory environment, in particular, has a marked influence on the beta of investment in electric utilities. However, betas in different countries reflect local factors and are not necessarily representative of those of other countries.

(iii)   Taxation and its relationship with the discount rates

Most of the previous discussion referred to post-taxation figures, since entrepreneurs are interested in their real returns after taxation. It is, therefore, better to apply the real required rate of return to post-tax rather than pre-tax cash flows. It is, however, necessary to calculate the relationship between pre-tax and post-tax required rate of return, since cash flows are usually presented in pre-taxation figures. This relationship depends on the depreciation policy of the firm (the annual writing down allowance, WDA), as well as the annual cash flows in proportion to the incremental initial investment. The relationship can be expressed by the following equation [7]:

$$\text{pre-tax return} = \text{post-tax return}/[1 - \text{Tax rate} \times (1 - \text{WDA}/p)]$$

where p = incremental annual operating cash flow expressed as the proportion of the incremental initial investment, or to put it in a simpler form:

$$\text{pre-tax return} = \text{post-tax return} \Bigg/ \left[1 - \text{Tax rate} \left(1 - \frac{\text{amount of depreciation}}{\text{annual cash flow}} \right) \right]$$

Therefore, the post-tax return depends on the tax rate, tax exemption procedures, the depreciation practices of the firm, as well as its annual cash flows. In all cases, pre-tax discount rates should be higher than post-tax discount rates. For instance, for a utility with a real post-tax return of 8 per cent, the required pre-tax return at a corporate taxation of 35 per cent can be as high as 12.3 per cent, but normally less than that after allowing for depreciation deductions.

Consider, for instance, an investment in a power station, which is supposed to cost £100 million and last for 20 years, where the discount rate is 10 per cent, and the tax rate is 30 per cent. Such a plant will need to have a net annual income to the investors after tax of £11.7 million to justify their investment (this is the annuity required to amortise an investment at a 10 per cent discount rate over 20 years). The pre-tax returns will have to equal the after-tax return plus the tax, after allowing for a tax allowance for depreciation (assuming a straight line depreciation of 5 per cent annually, i.e. 100 per cent divided by 20 years).

Pre-tax return = post-tax return (£11.7 million) + tax,

tax = (pre-tax return − depreciation) × tax rate of 0.3,

depreciation = £100 million/20 = £5 million,

substituting, the pre-tax return = £14.57 million.

Table 4.3   *Forecast cash flows in millions of dollars for the firm*

	Year 0	Years 1–20
Investment	100	
1. Revenue		300
2. Operational variable cost		250
3. Fixed costs		15
4. Depreciation		5
5. Pre-tax profit (1-2-3-4)		30
6. Tax		9
7. Net profit (5-6)		21
8. Operating cash flows (4 + 7)		26
Net cash flow	−100	26

Notes:
The assets are depreciated over 20 years in a straight-line method.
Tax rate is 30 per cent.
All figures of items 1–8 are annual figures.

Such a sum (from the Present Value Tables) corresponds to a pre-tax rate of return of 13.4 per cent, which is significantly higher than the expected post-tax opportunity cost of money of 10 per cent.

Consider another example that utilises financial statements. A firm that invested £100 million will have an expected annual cash flow forecast as in Table 4.3.

Whether the firm will proceed with the investment depends on the evaluation of a positive net present value of future cash flows at the firm's discount rate. Suppose that the firm's opportunity cost of capital (discount rate) equals 12 per cent after tax. Then a cash flow of £26 million, over 20 years, at a discount rate of 12 per cent will equal

$$\sum_{20} \frac{26}{(1 + 0.12)^n} = 26 \times \text{annuity factor}$$

$$= 26 \times 7.47$$

$$= £194.22 \text{ million}$$

(where 7.47 is the annuity factor obtained from the Present Value Tables).

The NPV of the investment is:

$$-100 + 194.22 = £94.22 \text{ million.}$$

Therefore, since the NPV is positive at a discount rate equal to the opportunity cost of capital, the firm will proceed with this investment.

(iv)   Inflation

All the above analysis was in real terms, i.e. ignoring inflation. However, if the cash flows are presented in nominal terms, i.e. in the money of the year in which they occur, then the discount rate has also to be in nominal terms.

Generally speaking, in the earlier defined CAPM model the *real* discount rate can be modified to *nominal* discount rate as follows:

real discount rate = real risk-free rate + (market risk premium × beta)

nominal discount rate = market risk-free rate + (market risk premium × beta).

Therefore, a utility operating in a utility market of beta 0.5, average market risk premium of 8.5 per cent and a nominal market risk-free rate of 8 per cent, will have a discount rate of:

$$8 \text{ per cent} + (8.5 \times 0.5) = 12.5 \text{ per cent.}$$

4.4   Controlling the value of the discount rate [6]

From the above discussion, and after allowing for risk-free returns plus premium for risk, inflation and taxation, the discount rate can assume high values [6]. This can affect investment in utilities like electric-power companies. It is, therefore, appropriate to consider ways of controlling the required rate of return for utilities. The most important means is through capital structuring and regulating prices (electricity tariffs).

4.4.1   Capital structuring

This involves raising a proportion of the capital required for investment through debt. Debt, for sound utilities, carries an interest, which can be as low as (or only slightly higher than) the market risk-free return on index-linked bonds, while the equity will require a high return because of risk. Therefore, capital structuring through having a significant part of the investment capital required in the form of fixed-interest loan can significantly reduce risk. However, it has to be realised that if there is a high debt/equity ratio then this entitles higher risk for the equity. Servicing the debt will be the first priority for the cash flow, which means putting returns to the equity capital at a higher risk. This means raising the beta of the equity, which may end in approximately the same rate of return for the total capital investment, irrespective of the debt/equity ratio.

Consider the case of a situation in which the real risk-free returns are 3.5 per cent, with risk premium on the average investment 10 per cent, totalling 13.5 per cent. If a power utility (with a beta of 0.6) is investing in a project totally by equity, its post-tax real returns should at least equal 3.5 per cent + 0.6 × 10 per cent = 9.5 per cent. If the capital is raised by a debt/equity ratio of 1:1, with a debt of real interest of 3.5 per cent, however, then the theoretical real cost of capital will become

0.5×3.5 per cent $+ 0.5 \times 9.5$ per cent $= 6.5$ per cent. This assumes that beta for the equity has remained constant. This is not the case, because the equity has become riskier owing to the need to service the debt, prior to having any returns to the equity holders. Beta may have to double to 1.2, so that the real return will remain roughly equal to the original 9.5 per cent.

Because of the tax deductibility of interest payments (*tax shield*), the main advantage of debt is the tax benefit. This may vary significantly from one country to another. Each country's situation and its taxation system have to be studied independently. Decision on the merits of tax deductibility of the interest rates and reduction of risk (if any) has to be taken into account when considering capital structuring.

4.4.2   Transferring risk

It is possible to transfer risk from the utility to consumers, by allowing continuous price (tariff reviews), which ensures that the equity holders will always get a decent pre-agreed target rate of return. This will almost eliminate the risk from the utility. A utility with regulated prices will have little risk and correspondingly a low beta and a low discount rate. This may tempt the utility to invest in higher capital-intensive assets, which will lead to higher costs to the utility and correspondingly higher tariff to consumers. This means shifting utility investment decision risks (whether right or wrong) to the consumers who have to pay, through the tariff, for the utility's decisions, i.e. the consumers now have to bear the additional risk of the investment to allow the utility always to get its target rate of return. Such an arrangement will not only shift the risk to the consumer, but will also prejudice the interest of other competing operators and investors in the industry who may not have the same privileges of price adjustments.

It is therefore essential when considering new projects and investments in regulated utilities to insist on using, as a discount rate, the private sector rate of return as employed in profitable tax paying investor-owned utilities.

4.5   Other forms of the discount rate

4.5.1   Weighted-average cost of capital

For project evaluation in the United States, most utilities use the *Revenue-Requirements Method (RRM)*. It is a project evaluation method that discounts future costs (revenue requirements) into their present value using the utility's *weighted-average cost of capital (WACC)*. WACC is the weighted-average cost of the firm's equity and debt. The least-cost alternative is the project alternative with the lowest present-value revenue requirement (PVRR), which is the lowest-cost alternative using the WACC as the discount rate [8].

This use of WACC is not consistent with methods of evaluation that are advocated by finance textbooks including this book. The WACC is an average between the cost of borrowing and the acceptable return on the investors' equity. Therefore, it reflects the overall cost of the utility's funds at any point in time. It is not the opportunity cost of capital and its utilisation as a discount rate is erroneous in this regard, because the discount rate for present valuing has to be equal to the opportunity cost of capital. The WACC does not reflect risk. Therefore, it is erroneous to utilise it for comparing projects of different risks – like comparing a nuclear plant (high-risk project) with a CCGT plant (low-risk project). The correct risk-adjusted discount rate for project alternatives is their opportunity cost of capital. As defined at the beginning of this chapter, in investment decisions, the opportunity cost of capital is the 'cut-off rate', below which it is not worthwhile to invest in the project. Normally the WACC is below the opportunity cost of capital and it correspondingly favours higher investment-cost project alternatives.

4.5.2   *Utilisation of multiple discount rates*

Different project alternatives have different risks. For instance, investing in a nuclear power station is riskier than investing in a conventional power station. New technologies have a much higher risk than established technologies. Therefore, for choosing the least-cost alternatives, more than one discount rate can be utilised, each for a different alternative. Each alternative has a different beta, which can be evaluated and incorporated in the CAPM model, to obtain a discount rate that is commensurate with the alternative's risk. That will allow for a risk-adjusted discount rate for each alternative, which assists in incorporating risk into evaluation of alternatives to choose the least-cost alternative and also to evaluate the required risk adjusted returns on investment. Risk assessment will be covered in detail in Chapters 13 and 14.

Multiple discount rates can be also utilised for different components of desegregated costs. Operational costs of a project contain many components: fuel, salaries, consumable and spare parts, local taxes and rates, insurance, etc. Each of these has a different future risk. For instance, local rents and rates can be fixed in real terms and need not be discounted with a discount rate that incorporates a risk element (beta is equal to 0). Fuel prices usually have much higher risk than any other cost parameter, particularly if there are prospects of environmental legislation that may entitle cleaner fuel versions (lower sulphur content, for instance) or plant modifications.

Such prospects are allowed for in the risk assessment. Alternatively, an inflation differential is incorporated in the fuel outlay, in the cost stream to cater for the possible increase in prices of fuel over the general inflation rate. For evaluation purposes, it is much easier to utilise a composite discount rate for the costs stream, while allowing for significant cost items (like fuel) to be treated in a manner that tallies with their possible variation from the general price increase trends.

4.6   Summary

The discount rate is very important in project evaluation and is a crucial factor in deciding the feasibility of a project. Investments in projects involve risk. Therefore entrepreneurs expect a post-tax real rate of return on their investment to equal the income of risk-free bonds plus a premium for risk. This premium depends on beta, which is the ratio of the fluctuation in the price of similar assets to the fluctuations in the overall stock market. In recent years, the real returns on government bonds averaged 2.6–3 per cent annually, while the average risk premium in the stock exchange, both in the UK and the US, was around 4–5 per cent [9].

The above post-tax rate of return has to be adjusted to a higher pre-tax rate of return. This takes into account taxation rates and exemption procedures, as well as the depreciation policies of the firm. If cash flows are in nominal, rather than real, terms, then a nominal discount rate that incorporates inflation has to be utilised.

Regulated utilities, like power companies, have a lower beta than the average equity. This depends on the type of business (generation or distribution), the type of facilities (conventional or nuclear), and the extent of regulation in the market. Such betas vary from one country to another and usually fall between 0.4 and 0.9.

The extent of risk can be slightly reduced by capital structuring, through increasing debt to equity ratio, and by transferring the risk to consumers, in a regulatory system that allows for regular electricity tariff reviews. This will only shift the investment risk from the firm to the consumers. It is, therefore, essential when evaluating investments in power utilities, whether public or private, that a rate of return (the discount rate) that is equal to that utilised in profitable tax paying investor-owned utilities should be employed.

4.7 References

1 MACKERRON, G.: 'Nuclear Costs: Why do they keep rising?', *Energy Policy*, 1992, **20**, pp. 641–652
2 'Electricity generating costs for plants to be commissioned in 2000' (UNIPEDE, Paris, January 1994)
3 CHRISTIAN D. J., and PRASAD, R.: 'Guidelines for calculating financial and economic rates of return for DFC projects' (International Bank for Reconstruction and Development, Washington DC, 1984)
4 DIMSON, E.: 'The discount rate for a power station', *Energy Economics*, July, 1989, **11**, (13)
5 'The economic appraisal of environmental projects and policies – a practical guide' (OECD, Paris, 1996)
6 BEALEY, R. A., and MYERS, S. C.: 'Principles of Corporate Finance' (McGraw Hill, 2000)
7 DIMSON, E., and STAUNTON M.: 'New Plant and Financial Factors', The review of the prospects of nuclear plant-consortium of opposing local authorities submission: volume 4, September 1994
8 AWEBBUCH, S.: 'The Surprising role of risk in utility integrated resource planning', *The Electricity J.*, April 1993
9 'Share valuations' (Financial Times, London, March 14th, 2003)

Chapter 5

Financial evaluation of projects

5.1 Introduction

5.1.1 Evaluation of costs and benefits

It is essential to recap on what was previously said in Chapter 3. During the life of the project, there will be two financial streams: one is the cost stream and the other is the benefits (income) stream. The two streams must contain all costs and benefits for the same estimated life frame of the project. In financial evaluation of small projects, the two streams will contain only the estimated actual cash costs and benefits of the project through its life cycle. The economical evaluation will influence the two streams, to include all the economic (social and environmental) costs and benefits of the project that can be evaluated. The difference between the two streams is the cash flow, the net benefits stream. The values of the net benefits can be negative, particularly during construction and the early years of the project. In later years, the benefits will usually exceed costs and the discounted net benefits will be positive, otherwise the project will not be undertaken.

It can be said that evaluation of costs and benefits of large projects (these normally demand equity, loans, and other financial instruments) is carried out on three levels [1].

Owner's evaluation

This is a straight evaluation. The owner is mainly concerned with cash flow and considers all money flowing in (income from sales, receipt of loans, and salvage value) as positive and all money flowing out (project cost, cost of operation, interest and repayment of loan) as negative. The owner is interested in net benefits and their net present value in comparison with the value of the investment (equity). The owner is concerned only with the return to equity.

Banker's evaluation

When analysing the project for loan consideration, the banker looks on it as one entity irrespective of the division of the investment cost between loans and equity. The banker evaluates the return on the total investment (equity plus loans) and considers its profitability. Therefore, the banker will consider the net present value of the whole investment and not just the investor's equity. The banker's evaluation integrates the points of view of the loan providers and equity investors.

Economic evaluation

This goes beyond the banker's evaluation to include all the economic (social and environmental) costs and benefits that can be evaluated. That includes taxes, subsidies, environmental and other external costs, as detailed later in Chapter 7. Such economic evaluation is usually done by development banks and similar institutions, and also by the concerned planning departments in the government.

5.1.2   Project financial costs

There are three main kinds of costs. Those are investment costs, operating costs, and working capital. Those costs are usually broken down into several different items as discussed below.

Investment costs

The items included under investment costs are (i) initial costs, (ii) replacement costs and (iii) residual values. Initial costs refer to those costs involved in construction and commissioning, including land, civil works, equipment and installations. Replacement costs refer to the costs of equipment and installations procured during the operating phase of the project, to maintain its original productive capacity. Residual values refer to the value of these investment items (equipment, land, etc.) at the end of the project's useful life. However, residual values are usually small and do not have a major impact on decision making. Sometimes, as in the case of nuclear power stations, decommissioning and dismounting represent a substantial cost, which can affect evaluation.

Operating costs

Operating costs are a combination of fixed and variable costs. Fixed costs will be incurred whatever the level of productions, like salaries, cost of management, and part of the maintenance cost. Variable costs will depend upon the level of production and include those items like fuel and energy, water, lubricants, and part of the maintenance cost (in the case of industrial projects, the cost of raw materials is also included). In general, production will never start at the maximum capacity of the project. Capacity utilisation may increase over time or may fluctuate according to time of day or to season. The variable cost will increase as production increases and will stabilise when maximum sustainable production capacity is reached. The total operating cost is the sum of the fixed and variable cost. The operating cost per unit of production falls as higher-capacity utilisation rates are achieved.

Working capital

Working capital refers to the physical stock needed to allow continuous production (spare parts, fuel, raw materials). The stock has to be built up at the commissioning phase and before the beginning of the commercial operation. Usually, stock requirements are defined as one month's worth of production. For power plants and network projects, there is usually one component of working capital, which is the initial stock of material necessary for commercial operation. But for industrial projects, there are three components of working capital: (i) initial stock of material, (ii) work in progress and (iii) final stock of production.

For evaluation purposes, all project costs are calculated at the estimated constant prices and costs (real terms) existing at a specific year – the base year.

5.1.3   Financial benefits of the project

Financial benefits of the project are brought about by selling the project product. These benefits are usually equal to the amount of production multiplied by the estimated base price. Not all projects in the electrical power industry imply production. Some, like efficiency improvement, lead to cost reduction, which is equal to the benefit. Others can have economic rather than financial benefit, as in the case of improvement of the supply reliability, rural electrification, and environmental preservation.

In the electrical power industry, calculation of benefits is not easy. A new power station would normally not only increase production, but also contribute towards reduction of the overall system cost of generation. It may also reduce system losses and delay the implementation of some projects for network strengthening. Certain projects are redundant and are made necessary by the need to ensure security of supply. Rural electrification is normally a source of financial loss, but has significant economic benefits. Some improvements in power stations, like inhibition of emissions, incur high investment, reduce electrical energy output and efficiency, and yet have sound economical (environmental) benefits.

Owing to deregulation, privatisation and competition, estimating financial benefits (profitability) of electrical power industry projects is becoming increasingly important. Later into the book, examples will be cited that help in understanding the methodology of estimating financial and economical benefits.

5.2   Least-cost solution

It is necessary to ensure that a project is profitable and that its return is higher than that of the opportunity cost of capital. Just as important, is to ensure that the project is the *least-cost alternative* for attaining the required output [2–4].

The ESI is one of the best venues for using least-cost solution techniques, since there is always more than one way in which a project can be executed so that its benefits can be secured. The least-cost solution aims at evaluating all realistic alternatives, financially and economically, before deciding the alternative that can achieve the project benefits at the least cost.

Projects in the ESI are, in the main, mutually exclusive; that is, the implementation of one project renders it technically or uneconomically feasible to implement another project.

Most day-to-day decisions in the electrical power industry involve financial evaluation to choose the least-cost solution (from mutually exclusive projects) for meeting the demand or rendering the service. For instance, there are many alternatives for meeting the need for more electricity generation: thermal power stations at different sites using small or large units firing coal or gas, gas-turbines and combined-cycle gas-turbines firing light fuel or natural gas, nuclear power stations, etc. Each alternative will have a different cost, cause a different system effect and lead to a new overall system cost. The least-cost solution aims at finding out the alternative technical arrangement that meets the requirements of electrical energy with the least cost to the utility, its site and timing.

The same applies to projects in the electricity network, sizing, timing of extensions, their routes and locations. Adjudication of offers for electricity facilities, like transformers where the high capital cost of an alternative can be traded against its lower losses, is required. The evaluation involves computing the overall cost of each network alternative (capital plus future cost of losses and other operational costs). Trading capital cost against future operational costs and against system security is a main criterion in financial project evaluation and least-cost solution. Therefore, in most electrical power engineering decisions, there is more than one alternative to achieving the required result. The least-cost solution considers all these alternatives, evaluates them and indicates the alternative with the least discounted overall cost over the useful life span of the project. Appendix 1 details an example for evaluating the least-cost alternative in choosing a transformer while trading capital cost against cost of losses.

In choosing the least-cost solution, we are concerned with the differences in the present value of the cost of the alternatives (including their system effects if any). In many cases, the benefits (output) of each alternative are the same since all the alternatives are supposed fully to meet the project. In this case we are concerned with the evaluation and comparison of costs. However, if there are differences in the amount of the energy output, then comparison of alternatives is carried out per discounted kWh of electricity output, through evaluating over time the overall system cost of the alternatives.

There are many methods for financial evaluation and comparing alternatives. The most important and useful ones are:

- the present value method, and
- the annual cost method (the equivalent uniform annual cost method).

5.2.1   Present value method

The present value (PV) method [5,6] aims at present valuing (discounting) all costs and benefits of the project or cash flows (net benefits) to a specified date, the 'base year'. In this case, all cash flows prior to or after the base year are discounted to the base year through multiplying by the discount factor $[1/(1 + r)^n]$ where 'n' is negative for years prior to the base year. All values are considered to occur at the year

Table 5.1   Present valuing (all values are in thousand pounds)

Year	Costs			Income	Net benefits		
	Capital	Operational	Total		Cash flows	DR = 8%	DR = 15%
−1.50	100	—	100	—	−100	−112.24	−123.32
−0.50	50	—	50	—	−50	−51.96	−53.62
0	Commissioning date						
0.5	20	20	40	40	0	—	—
1.50	—	30	30	80	50	44.55	40.54
2.50	—	30	30	80	50	41.25	35.26
3.50	—	30	30	80	50	38.19	30.66
4.50	—	30	30	80	50	35.36	26.66
5	Residual value			60	60	40.83	29.83
				Present value of net benefits:		35.98	−13.99

end. However, in most instances, the middle of the year may be the more appropriate date for their aggregation, as already discussed in Chapter 3.

To give a simple example of aggregating flows at the middle of the year, consider a project that is supposed to live for five years after completion, is expected to have the following two streams of cost and income based on the base-year prices, and at two discount rates of 8 per cent and 15 per cent. The project will have a residual value of £60 000 at the end of its useful life of five years.

The example in Table 5.1 demonstrates the present valuing procedure. Costs and benefits are supposed to be evenly distributed over the appropriate months of the years of the project. Therefore, these can be grouped in the middle of the year about the commissioning date (base year). This project will have a net positive value of £36.07 thousand, at a discount rate of 8 per cent, but a negative value of £13.99 thousand, at a discount rate of 15 per cent. Correspondingly, the project is profitable if an opportunity cost of capital of 8 per cent is considered appropriate. The project needs to be discarded if 15 per cent is the opportunity cost of capital. It is also possible to avoid the discounting for the year plus half by choosing the base date to occur yearly from the middle of each accounting year, as shown by the example in Appendix 1.

The PV method is suitable for choosing the least-cost solution. The method discounts the capital and future running costs of each considered alternative to its present value, using the agreed discount rate that is the opportunity cost of capital. If there is any salvage value for the alternative, its value after discounting to the base year is deducted from the cost of this alternative. Different alternatives may have different system effects, and correspondingly different costs. These have to be undertaken in the evaluation. The alternative with the least present value cost, is the chosen least-cost solution.

Table 5.2   Comparing alternatives

Year	Alternative 1				Alternative 2			
	Cost (£)		Production (kWh)		Cost £		Production (kWh)	
	Actual	Discount	Actual	Discount	Actual	Discount	Actual	Discount
1					10	11	0	0
0	100	100	0	0	150	150	0	0
1	50	45.45	500	454.5	50	45.45	500	454.5
2	30	24.79	1000	826.4	10	8.26	1000	826.4
3	30	22.45	1000	751.3	10	7.51	1000	751.3
4	30	20.94	1000	683	10	6.83	1000	683
5					10	6.21	1000	621
Sum	240	213.63	3500	2715.2	250	235.26	4500	3336.2

Cost per discounted kWh, for alternative 1 = £213.63/2715.2 = £0.0787; for alternative 2 = £235.26/3336.2 = £0.0705. (As explained earlier the base date was chosen so that the annual costs and benefits grouped at the middle of each year are at yearly intervals from the base date.)

In choosing the least-cost alternative, benefits need not be presented in monetary terms. In some cases, benefits are better presented with the physical output of the project such as number of units produced, tons of the output or any similar output. This is particularly suitable in the case of an electrical power project, since most of the benefits are production of electricity as kWh, its multiples, or reduction of cost as reduced losses or usage in kWh. The method is also particularly useful for comparing alternatives with different lengths of execution time and different outputs (kWh). The evaluated costs (capital and operational) and benefits (in kWh) are discounted to the base year. The least-cost alternative will be the one least discounted cost divided by the discounted energy output. This is demonstrated using a 10 per cent discount rate in Table 5.2. Alternative 1 is the least-cost solution.

5.2.2   Annual cost method (equivalent uniform annual cost method)

This is a useful and quick way for choosing the least-cost solution. It can help in providing the right answer, supposing that certain assumptions and approximations are possible.

In Chapter 3, the equivalent annual cost was defined as being equal to the amount of money to be paid at the end of each year 'annuity' to recover (amortise) the investment, at a rate of discount 'r' over 'n' years. An annuity is a level stream of cash flows that will continue for a specified number of time periods (years). The annuity factor is the present value of all the annual values of the discount factors over the whole period (annuity factor $= \sum_N$ discount factor).

The equivalent annual cost 'M' will be equal to:

$$\frac{\text{present value of the investment}}{\text{annuity factor}},$$

where annuity factor $= \left[1/r - 1/(r(1+r)^n)\right]$.

The same results can be obtained by multiplying the present value of the investment by the capital recovery factor (CRF), since CRF is equal to [1/annuity factor].

Alternatively, the annual benefits (B) of the project, if they are equal throughout the years of the project, can be present valued by the same way:

$$\text{present value of benefits} = \text{annual benefits (B)} \times \text{annuity factor.}$$

The same applies to present value the annual fixed operating cost (FC) and annual running operational costs (OC), throughout the life of the project. If these are constant every year (for a known output), these can be present valued as:

$$\text{annuity factor} \times (FC + OC).$$

Therefore, the present value of the total costs of the project throughout its useful life will equal to:

$$\text{project cost at base year} + [(FC + OC) \times \text{annuity factor}]$$

and the total benefits of the project $= B \times$ annuity factor.

For electricity generation projects, such benefits can be in the form of kWh. If benefits are in monetary terms then annual cash flows (net benefits) are utilised.

The PV of a project with equal annual operational cash flows is equivalent to:

$$\text{project cost at base year} + (\text{annual cash flow} \times \text{annuity factor}).$$

(It has to be remembered that project cost is negative because it is an outward-flowing payment.)

This is a useful and quick way for comparing alternative projects, and for approximately calculating the cost of production and prices. A good example is to compare the cost of production from a power station utilising a combined-cycle gas-turbine plant (CCGT) with that utilising steam-power turbines (ST) of a similar size. Installed cost at commissioning of the CCGT plant is £500 kW^{-1}, and it has an expected life of 20 years. The ST plant costs £1000 kW^{-1} with an expected life of 30 years. The fixed and running costs (fuel, operation and maintenance) are also 2.4p kWh^{-1} and 2p kWh^{-1}, respectively. Each plant will operate at full load; approximately 8000 h annually for ST and 7000 h annually for CCGT. All these costs apply to the same commissioning date.

Then, the production cost of each kWh from these two alternatives, with a discount rate of 10 per cent, can be calculated as follows. The annuity factors for 20 and 30 years at 10 per cent discount rate are 8.514 and 9.427, respectively.

$$\text{For CCGT: } [(£500 \div 8.514)/7000] + 2.4\text{p} = 3.24\text{p kWh}^{-1};$$

$$\text{For ST: } [(£1000 \div 9.427)/8000] + 2\text{p} = 3.33\text{p kWh}^{-1}.$$

Obviously, production from the CCGT power station is the least-cost alternative. The above calculation can also assist in the assessment of the tariff to be charged for such a production. An alternate way of calculation is to present value all the future running cost of the alternatives, to present value the output in kWh (as benefits), and make the necessary comparison. Therefore, for the CCGT alternative (per kW of installed capacity):

$$\text{PV of operational cost} = 0.024\text{p} \times 7000\,\text{kWh} \times 8.514 = £1430.4,$$

$$\text{Total PV of costs per kW} = £500 + £1430.4 = £1930.4,$$

$$\text{PV of benefits (kWh)} = 7000 \times 8.514 = 59\,598\,\text{kWh},$$

$$\text{Cost per kWh} = £1930.4/59\,598 = 3.24\text{p}.$$

For the ST alternative:

$$\text{PV of operational cost} = 0.02 \times 8000\,\text{kWh} \times 9.427 = £1508.3,$$

$$\text{Total PV of costs per kW} = £1000 + £1508.3 = £2508.3,$$

$$\text{PV of benefits (kWh)} = 8000 \times 9.427 = 75\,416\,\text{kWh},$$

$$\text{Cost per kWh} = £2508.3/75\,416 = 3.33\text{p}.$$

This alternative method of assessment yields the same result as above.

This is a quick method that gives rapid results and allows the attention of the evaluator to focus on a few alternatives. However, it involves many assumptions and approximations that need to be handled carefully. Therefore, such quick methods have to be elaborated further to consider other detailed issues (exact commissioning, system effects, environmental impacts, long-term prospects, etc.) to allow accuracy in decision making.

The choice of either the PV or the annual cost method for evaluating the least-cost alternative is a matter of personal convenience. In the United States, the annual cost method is preferred. Its implications are easier to understand for business decisions, and it is easier to compute for regular annual series of disbursements; particularly if the capital is obtained through loans. In the electrical power industry, annual costs are irregular, utilisation varies from one year to another, and most decisions affect the overall system cost. To allow for more detailed evaluation of future costs, the PV method is preferred for identifying the least-cost solution since it allows for variations in output, operating costs over time and other factors.

5.3   Measuring worth of the investment

Many of the projects in the electrical power industry are forced on the planner and are beyond his control; examples are building new power plants to meet load increase, providing new facilities to enhance security of supply, rural electrification, etc. There-fore, many decisions in the electrical power industry are restricted to the choice of the least-cost solution. With the privatisation of utilities and market economies, a more thorough analysis of the profitability of investment is also becoming essential. Projects are carried out because they are needed, they are the least-cost solution, and they are profitable. It is now becoming increasingly necessary not only to weigh the benefits of the investment through a cost–benefit analysis, but also to carry out financial prof-itability projections. This includes forecasting three summary statements: (i) the *pro forma* income statement, (ii) the *pro forma* balance sheet and (iii) the *pro forma* funds flow statement [7]. Such statements will allow assessment of financial profitability to owners.

In this section, we are concerned with the traditional cost–benefit analysis of projects to assess their acceptability to utilities, governments, investment bankers and development funds. There are several ways of assessing whether the project is worth undertaking. The most useful of these are:

  (i)   computing the internal rate of return
 (ii)   evaluating the net present value of the project
(iii)   calculating the benefit/cost ratio
 (iv)   other criteria (pay back period, profit/investment ratio, commercial return on equity capital).

All the above criteria, except for the last, involve discounting.

5.3.1   Internal rate of return

Calculating the internal rate of return is a popular and widely used method in the evaluation of projects. The internal rate of return (IRR) is the discount rate that equates the two streams of costs and benefits of the project. Alternately, it is the rate of return 'r' that the project is going to generate provided the stream of costs (C_n) and stream of benefits (B_n) of the project materialises. It is also the rate, r, that would make the NPV of the project equal zero, i.e. IRR is such that:

$$\sum [C_n/(1+r)^n] = \sum [B_n/(1+r)^n].$$

If IRR is equal to or above the opportunity cost for a private project, or the social discount rate (as set by the government) in public projects, then the project is deemed worthwhile undertaking. Utilities, governments and development funds set their own criteria for the opportunity cost of capital and for the social discount rate below which they will not consider providing funds. Such criteria depend on the amount and availability of required funds. Criteria also depend on the presence and expected return of alternative projects in other sectors of the economy, the market rate of

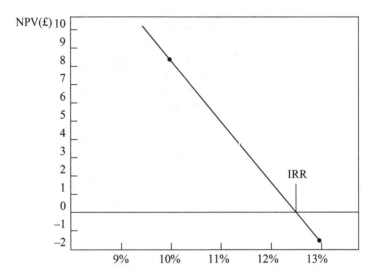

Figure 5.1   Calculating the IRR by interpolation

interest, and the risk of the project. In Chapter 4, the concepts that govern the choice and fixing of the discount rate were discussed. For some investors, the discount rate can be viewed as the minimum acceptable rate of return below which a project is rejected. Therefore, if the IRR is equal to or above the minimum acceptable rate of return, then the project is considered to be worthwhile undertaking. The IRR can be calculated by trial and error calculations, through the utilisation of the above equation until r is found or can be interpolated. However, preferably, it can be calculated through the use of a computer program or a modern financial calculating machine.

Alternately, it can be computed using a graphical method. Two points may be plotted on graph (Figure 5.1) and joined by a straight line. The point at which this line cuts the horizontal axis (i.e. where the NPV is zero) gives the IRR. The IRR for the example in Table 5.3 is shown.

The IRR concept has certain minor weaknesses that have been explained in the literature [8,9] and can sometimes be defective as a measure of the relative merits of mutually exclusive projects. It also contains an important underlying assumption, that all recovered funds can be reinvested at an interest rate equal to the IRR, which is not always possible. However, the IRR is a widely understood concept and it largely represents the expected financial and economical returns of the project. Also most of the weaknesses referred to do not normally occur in the electrical power industry. The main merit of the IRR is that it is an attribute of the project evaluation. Its calculation does not involve an estimation of a discount rate. Therefore, the evaluator avoids the tedious analysis of Chapter 4. It is satisfactory to calculate the IRR and compare it with the test rate conceived, which is a superficial attractiveness. Therefore, it is a widely used means of assessing the return of the project in the electricity supply industry.

Table 5.3   Calculating benefits

Year	Cost	Income	Net benefits	Net benefits discounted at 10%
−1	40	—	−40	−44
0	100 + 10	40	−70	−70.00
1	10	40	30	27.27
2	10	40	30	24.79
3	10	40	30	22.54
4	—	70	70	47.81
			Net present value: £8.41	

5.3.2   Net present value (NPV)

The concept of net present value has already been fully explained. The method discounts the net benefits (cash flows), i.e. project income minus project costs to their present value through the already assigned discount rate. If the net result is higher than zero, this proves that the project will provide benefits higher than the discount rate and is worthwhile undertaking.

For example: consider the same project in Table 3.1, with the following present-day following costs (investment £140, annual running cost £10) and income (benefits) of £40. The project is expected to last four years and the prevailing discount rate of 10 per cent is used.

Calculating the IRR: since the NPV at 10 per cent is positive, then the discount rate is greater than 10 per cent. Trying 12 per cent, the NPV is £1.76; at 13 per cent the NPV $= -1.47$. The IRR is, therefore, around 12.5 per cent. Since it is higher than the minimum acceptable return of 10 per cent, the project is acceptable.

Finally, we consider the NPV: since the NPV > 0, the project is acceptable.

The method of the NPV is a powerful indicator of the viability of projects. However, it has its weaknesses in that it does not relate the net benefit gained to the capital investment and to the time taken to achieve it. In the above example, it does not matter if the £8.41 were obtained through investing £100 or £1000, or obtained over 4 years or 40 years. However, it is very useful in choosing the least-cost solution, since it is the alternative that fulfils the exact project requirements and has the higher NPV that is preferred.

5.3.3   Benefit/cost ratio

This method compares the discounted total benefits of the project to its discounted costs:

$$B/C = \sum_n B/(1+r)^n \Big/ \sum_n C/(1+r)^n.$$

Only projects of $B/C \geq 1$ are adopted. This criterion is popular and, in some applications, is more useful than the net present value, in that it relates the benefits to costs of the projects. It also has a useful application for capital constraint, when the industry has a lot of feasible projects but limited investment budget. In this case projects are ranked in accordance with their B/C ratio and are adopted accordingly until their combined costs equal the capital investment budget. Therefore, it is useful in comparative analysis.

It has to be remembered that B/C is a ratio, while NPV is measure of absolute value.

In the example of Table 5.3:

value of discounted benefits are $= 40 + 36.36 + 33.06 + 30.05 + 47.81 = £187.28$,

value of discounted costs are $= 44 + 110 + 9.09 + 8.26 + 7.51 = £178.86$,

$$B/C \text{ ratio} = 187.28/178.86 = 1.047.$$

5.3.4   Other non-discount criteria for evaluation

There are many other criteria for evaluating the project's worth and return of investments. The most common methods are the payback period and profit/investment ratio calculation.

5.3.5   Payback period

The payback period is defined as the time after initial investment until accumulated net revenues equal the investment, i.e. the length of time required to get investment capital back. The method is used as an approximate measure of the rate at which cash flows is generated early in the project life. It is utilised for small investments, like improvements and energy efficiency measures, since it is easy to understand by business managers. However, in isolation, it tells the analyst nothing about the project earning rate after payback and does not consider the total profitability or size of the project. It also ignores inflation and discriminates against large capital-intensive infrastructure projects with long gestation times. Therefore, it is a poor criterion in itself and it must be used in conjunction with other criteria.

In the previous example of Table 5.3 the payback period is higher than three years.

5.3.6   Profit/investment ratio

The profit/investment ratio is the ratio of the project's total net income to total investment. It describes the amount of the profit generated per pound invested, and is sometimes referred to as profitability index. The idea is the selection of the project that maximises the profit per unit of investment. The ratio is easy to calculate, but does not reflect the timing at which revenues are received and profits generated. Therefore, it does not reflect the time value of money. Correspondingly, it is not a proper evaluation measure. In the example of Table 5.3 the profit/investment ratio is $£50/£140 = 35.7$ per cent.

5.4 Owner's evaluation of profitability – commercial annual rate of return

At the beginning of Section 5.3 it was explained that, for an investor to assess commercial profitability, forecasting a *pro forma* income statement through conventional financial accounts is necessary.

Therefore, it is possible roughly to evaluate the profitability of the equity investment from the commercial viewpoint of the investor through calculating the equity's profit by financial accounts. A year is chosen at which the project matures, and the net profit of the project is calculated for that year. Net profit is equal to income less operational expenses, depreciation, interest, taxes, and other expenses. This is compared with the total share (equity) capital invested and is the commercial return to the investors.

The same approach can be utilised for assessing the return on the total investment (equity plus loans). In this case the net profit, for commercial purposes, will be equal to income less operational expenses, depreciation, taxes, and other expenses. Interest is not included, since interest is the return on the loans. This net profit is divided by the total investment to calculate the commercial return of the whole investment.

Such commercial accounts are not the criteria for proper project evaluation for several reasons; mainly because they do not account for the time value of money. Depreciation is used as cost of capital invested, and depreciation does not adequately account for the declining value of money or for inflation. Accountants' measures of profitability are not well suited for *ex ante* project evaluation, especially for non-commercial projects.

Evaluation methods that do not involve the time value of money (discounting) are poor indicators. They provide approximations and should not be utilised for least-cost solutions, or evaluating large investments.

To summarise: the guiding principle for project evaluation is the maximisation of net present value while utilising, as a discount rate, the opportunity cost of capital. The internal rate of return is not the only criterion for evaluating projects for investment decisions. Net present value with a proper discount rate (reflecting the true opportunity cost of capital) is a criterion. With limited budgeting a benefit/cost ratio has to be calculated to assist in prioritising projects.

5.5 References

1 'Discounting and Measures of Project Worth', University Of Bradford, Development and Project Planning Centre, 1995
2 STOLL, H. G.: 'Least-cost, electric utility planning' (John Wiley and Sons Inc., 1989)
3 TURVEY, R.: 'Optimal pricing and investment in electricity supply' (George Allen & Unwin Ltd., 1968)
4 BERRIE, T. W.: 'The economics of system planning in bulk electricity supply' *in* TURVEY, R.: 'Public enterprise' (Penguin Book, 1968)

5   NORRIS, T. E.: 'Economic comparisons in planning for electricity supply', *Proc. IEE*, March 1970, **117**, (3)
6   BREALEY, R., and MYERS, S.: 'Principles of corporate finance' (McGraw-Hill Inc., 2000)
7   DUVIGNEAU, J. C., and PRASAD, R. N.: 'Guidelines for Calculating Financial and Economic rates of return for DFC projects', World Bank technical paper 33
8   LEAUTAND, J. L.: 'On the fundamentals of economic evaluation', *Eng. Econ.*, **19**, (2)
9   JEYNES, P.: 'Fundamentals of accounting/economic/financial evaluation', *Eng. Econ.*, **20**, (4)

Chapter 6

Considerations in project evaluation

Apart from discounting and least-cost evaluation, there are many considerations in project evaluation. These include: allowing for contingencies, timing of expenditures, sunk costs, depreciation and interest charges. There are also other considerations relating to system linkages, dealing with projects of different lives, and expansion projects. Such considerations are discussed in the following sections.

6.1 Base cost estimate and contingencies

The base cost estimate [1] done at the evaluation stage is the best judgment of the cost of all project components at the specified evaluation date. It assumes that these components are accurately known and properly costed. If certain cost estimates are rough, then that indicates the need for further investigations to assess accurately project cost components. The period for project implementation and the commissioning date should be estimated.

The best estimates of project costs can be undertaken from prices available at the time of estimation (date of evaluation). Project implementation can take many years. Hence, the actual cost of the project at the commissioning date will differ from the cost estimated at the time of planning and evaluation. This is the result of inflation and increase in prices and quantities of the project components. Such variation is allowed for through the following two contingencies.

(i)   A physical contingency to allow for the increase in quantities of material and equipment involved in project execution and the change of implementation methods.
(ii) A price contingency to allow for increase in price of project components, over the base case estimates, during the construction period.

6.1.1   Physical contingency

The physical contingency reflects possible changes in quantities and implementation procedures. The amount of physical contingency depends on the type of the project components. Civil works have a higher contingency than machinery, to the extent of detailing in project estimate preparations. Physical contingencies usually vary up to 10 per cent of the project base cost. Beyond this, a more detailed estimate has to be undertaken in order to reduce uncertainties in project cost estimation. Physical contingencies are distributed along the project's execution life as a percentage of annual cost allocations.

6.1.2   Price contingency

The price contingency allows for the increase in unit prices of the project components, beyond the estimated prices of the planning year. It is highly dependent on inflation rates and other possible price increases. This allows estimating project cost disbursements in the money of the year at which they occur. The sum of these annual disbursements including price and physical contingency give an idea of project cost for formulating a *finance plan*. The full finance plan will include these costs plus the interest during construction and working capital. This indicates the amount of money needed to be raised and procured to finance the project. However, the project cost for financial and economical evaluation purposes is not indicated, since the financing plan involves adding money of different dates and correspondingly different real values. The base project cost is that project cost estimated in the money of the planning year (or the commissioning year).

If the project cost is distributed along the years of the execution period and is desired to utilise the commissioning date as the base year and to present the project cost in the money of that year, then the anticipated annual disbursements, during the project execution period (which include both physical and price contingencies), are compounded to that year by the nominal discount factor (which includes the discount rate plus inflation). The example of Table 6.1 assists in explaining this. An alternative arrangement, if the project is estimated as a lump sum in the money of the planning year and it is desired to have the commissioning date as the base year, is where the planning year estimates are compounded to the commissioning date by the nominal discount rate with the physical contingency added, and these project costs are then represented in the money of the year of commissioning.

For example, consider a project that takes three years for execution, and costs £100 at the base year (the present year of planning). Investment is grouped to be disbursed

Table 6.1   Financing plan

Year	Base	Physical contingency	Price contingency	Total
1	20	2	1.1	23.1
2	40	4	4.5	48.5
3	40	4	6.9	50.9
(commissioning)				
		Total project cost for the financing plan purposes = £122.5		

at 20, 40 and 40 per cent spread over three years starting one year after the cost estimation date. Price contingency is equal to inflation at 5 per cent annually. Physical contingency is 10 per cent and the discount rate is 12 per cent. Commissioning date is after three years. The financing plan requirements are in Table 6.1.

The project cost in the money of the base year is equal to the base cost plus the physical contingency (i.e. £100 + £10). The project cost at the commissioning date is the base cost plus the physical contingency, compounded by a factor of 1.626, i.e. $(1.12 \times 1.05)^3$, where (1.12×1.05) is nominal discount rate. Therefore,

project cost at the base (planning year) = £110 (money of the base year),
project cost at the commissioning year = £178.86 (money of the commissioning year),
project financing requirements = £122.50 (plus interest during construction and working capital if any).

This is elaborated further in Section 6.2.

6.2   Interest during construction

An important consideration in project evaluation is how to account for interest during construction. In calculating the IRR of the project, the project cost can be considered to be equal to a single evaluated cost at the base year. If this is done, then it is not necessary to account for inflation or interest during construction in calculating the IRR. If the base year is the project execution commencement year, then the project cost at that year is the estimated cost of the project in the money of that year plus the physical contingency. If the project is costed in the money of the commissioning year, then the price contingency is added to the cost stream of payments for project execution, so as to render it in the money of the respective price contingency year. Then this stream is discounted (compounded) to the commissioning year by the nominal discount (compounding) factor, as already explained in Section 6.1. Usually, the nominal discount rate is not much higher than the interest rate utilised for financing construction (interest during construction). The combined price contingency and the discount (compound) rate account for the interest during construction, where the latter is ignored in evaluation.

Therefore, for financial and economical evaluation of projects, the project capital costs can be presented as a lump sum either in the money of the project planning year (commencement of execution) or the project commissioning (completion) of execution year. The latter figure is obtained by compounding the actual annual disbursements on the project to the commissioning year through the nominal discount rate. Alternatively, the project cost, while estimated in the planning year, can be presented as stream of payments along the execution period. In this case the physical and price contingencies are added and this stream is discounted by the nominal discount rate to the base year, whether it is the year of planning or commissioning.

Therefore, interest during construction is not directly considered in the financial and economical evaluation. It is ignored if the above procedures are carried out correctly. However, interest during construction is important for planning the finance of the project. It has to be realised that if the project is financed entirely by equity, then there will be no payments for interest during construction.

6.3   Sunk costs

Sunk costs are defined as those costs that have been incurred on the project, before commencing the present evaluation. Despite their value to the project, they have already been spent. Therefore, they are not included in the costs of the project for new decision making or for determining whether to proceed with the project or not [2]. An investment is sunk if the asset cannot easily be used for any purpose other than its original purpose. Sunk costs can be different from fixed costs of a project. For instance, an investment in a power station building is sunk because the building has no other use (without great cost and difficulty). On the other hand, an investment in building management offices for a project in the city (although it is a fixed cost) is not a sunk cost, since such an asset can be readily converted for other purposes [3].

When evaluating the costs and benefits of completing or extending a project, the costs and benefits that already exist are not included in the evaluation. It is only the expected future costs and benefits that matter in the decision making today. Therefore, the IRR of completing unfinished projects or extending existing projects would appear to be very high, but that reflects the nature of the decision being made.

So if it is decided to convert or extend an existing installation or to utilise an existing infrastructure, there is no need to include past investments in the new evaluation. These are sunk costs, which should not be considered. However, it may be appropriate sometimes to calculate the returns on the total project (including sunk costs) to evaluate earlier decisions and learn of such experiences. This, however, need not interfere with the present evaluation decision making.

Sunk costs occur frequently in the electricity supply industry, for example in a decision to extend or refurbish an existing power station or substation site, updating an existing transmission line, converting an existing dam into a hydro-power facility, etc. In all such cases, it is only the added costs and benefits of the project that need consideration. In evaluating the original project, the prospects of future extensions and modifications are contemplated. Such prospects should be included in the evaluation of the original project, since the success of the original project evaluation may be contingent on such future prospects.

6.4   Depreciation and interest charges

In evaluation, the cost and benefit streams of the project are reduced to cash flows. As already defined, cash flows are the difference between pounds received and spent out. Therefore, in calculating the internal rate of return, depreciation and interest charges are ignored. Depreciation does not represent actual cash flow and correspondingly is not involved in present valuing or any other economic evaluation criteria.

Depreciation is, however, important in the preparation of financial statements (financial projections) of private utilities and firms. It influences the taxation, the profit/loss statement, and hence dividends to shareholders.

With regard to the interest charges, it must be remembered that the IRR is the maximum rate of interest that could be paid for the funds employed over the life of the investment without loss to the project. Therefore, interest charges also should be excluded from calculating the rate of return of the project and other evaluation criteria [3,4]. This equally applies to interest during construction.

In most projects, financing is done by a combination of equity and loans. The investor is interested to know the IRR of the project and the return on equity. In calculating the return on the equity, the amount of loan servicing (interest on the loan and loan repayments) has to appear as cash expenses and deducted in the appropriate years from the current income stream. Simultaneously, the amount of loan has to be deducted from the project cost. Thus, the costs, benefits and net benefits will represent those accruing to the equity only. The IRR calculated will be the rate of return on the equity. It has to be compared with the opportunity cost of the capital, i.e. the return that this equity capital can obtain in the best alternative investment. The investor would undertake the project only if the internal rate of return on the equity in this project is higher than the opportunity cost.

6.5 Financial projections

It is useful to carry out financial projections for a new project. Such financial projections help in calculating the cost and benefit streams and consequently computing the project's IRR. The financial projections [5] necessary are: (a) projected income statement, (b) projected balance sheet and (c) flow of funds statement. The income statement is useful in project evaluation, since it contains most of the information needed for the financial and economical evaluation. The income statement (in nominal terms) would detail for each year the income of sales and services, the cost of products (fixed plus operation), maintenance and capital charges (depreciation and interest). The information will allow the calculation of the profits in any given year and, after deduction of taxes, the calculation of net profits for distribution to the equity owners.

The financial statement is produced in real terms; usually in the figures of the base year. It is more appropriate to plan the financial projections in current (nominal) terms to ensure coverage of full capital costs by the finance plan. It is also appropriate to evaluate the financial development of the firm and the viability of the project under the realistic situation of inflation. Therefore, the real financial projections (of the base year costs and revenues) are inflated, year by year, with a figure normally equal to the expected inflation rate. This will result in financial projections in current (nominal) terms. Such current projections will enable one to ascertain that the financing available in the form of equity, loans, financial allocations (depreciation) and retained profits will be sufficient to meet financial requirements of the project in current terms, year by year.

A single inflation rate may be useful for the estimation of financial statements. In some cases, and as already mentioned, certain items of costs and benefits may deviate from the expected general inflation level. For instance, expected increases in fuel prices may be higher than the inflation rate. Also, electricity price increases may display a trend of being slightly lower than the general price increase. In such cases, inflation differentials should be included in drawing out the financial statement to allow more realistic assessment of future trends. Such inflation differentials are also important when drawing out the cost and benefit streams in real terms for calculating the project's IRR. For instance, if it is expected that fuel prices, in the future, will increase at y per cent annually over the general inflation rate, then the real outlay for fuel in the cost stream should increase at a compound rate of y per cent annually over the base year.

The income statement, in current terms, aids in calculating the tax on profits. Most private investors are interested in the *after-tax financial rate of return*. Therefore, the income statement will allow the compilation of tax for each year. This is deducted from the net benefit stream to allow computation of the after-tax financial rate of return. Also, in the case of existence of loans, the deduction of the loan component and loan servicing from the benefit stream will enable the calculation of the after-tax financial rate of return on equity.

It is, however, essential to understand the concepts of financial statements: income statement, balance sheet and funds flow statement, in order to assess the actual performance of the company (firm) in the commercial market and the profits that will be paid to shareholders. It was explained in Section 5.4 that such financial statements (commercial accounts) are not criteria for proper project evaluation or least-cost solution. However, preparation of *pro forma* financial statements and financial projections help not only in computing the cost and benefit streams and consequently the IRR and NPV, they also project the financial and commercial profitability of the project and the actual financial return that is likely to be paid to owners and share holders. In case of projects, analysing financial performance and the calculation of financial ratios is useful fully to grasp the envisaged performance of the firm, and impact of the project on its financial performance and profitability to owners, the financial problems the firm is likely to face (availability of funds, debt servicing, liquidity problems, etc.) as well as to undertake comparisons of the firm's performance with market acceptable financial ratios. Such financial performance and financial ratios will affect the ability of the firm to raise funds in the money market and the terms of these, as well as the price of the firm's share if these are traded in the stock exchange. Sections 6.5.1–6.5.3 offer short explanations of financial statements and ratios.

6.5.1   Financial performance and financial statements

Typical financial statements [6,7] will look like those of Table 6.2, which contains an income statement followed by a balance sheet. These are summary financial statements. Annual reports of companies will show these summary financial statements as well as notes that explain, in detailed tables, each of the items that appear in the summary financial statements, including a table for allocation of profits (if any) which

Table 6.2   Summary pro forma *financial statement for a power generating company (figures are only indicative and in £ million)*

	2002	2003
Income statement		
Income of electricity sales	125.0	128.8
Other income	2.0	2.7
Costs of generation (other than fuel)	18.0	18.5
Fuel costs	50.0	52.0
Other expenses	8.0	7.0
Depreciation	25.0	25.0
Earning before interest and tax (EBIT)	26.0	29.0
Net interest	10.0	9.0
Tax at 40%	6.4	8.0
Net income	9.6	11.7
Preferred stock dividend (if any)	—	—
Earnings of the common stock	9.6	12.0
Balance sheet		
Cash and short-term securities	5.5	5.0
Receivables	20.0	21.0
Inventories	15.0	16.0
Other current assets	2.5	2.5
Total current assets	43.0	44.5
Land, plant and equipment	554.0	544.0
Other long-term assets	20.0	40.0
Total assets	617.0	628.5
Debt due	10.0	10.0
Payables	100.0	95.0
Other current liabilities	5.0	5.0
Total current liabilities	115.0	110.0
Long-term debt	100.0	90.0
Other long-term liabilities	102.0	118.9
Preferred stock (if any)	—	—
Shareholders equity	300.0	309.6
Total liabilities	617.0	628.5

explains how profits applicable to common stock will be divided between dividends, reserves and retained profits.

The income statement of Table 6.2 can be a *pro forma* income statement for a power-generating firm that is supposed to commence its full production early in 2002; the figures of 2003 are also *pro forma* figures based on same energy sales, with a higher income of 3 per cent, which is the general inflation index expected for that year. However, prices of fuel have been inflated by 4 per cent in accordance with agreements and general price trend for fuels.

In the income statement, the income from electricity sales as well as other income (which can be from selling services or income from other investments) represents the total income of the firm. From this, costs are deducted; these are costs of operation (salaries, consumables, etc.) as well as fuel cost and other expenses (rents, rates, etc.). The allocations for depreciation are also deducted. The result is earnings before interest and tax (EBIT). After deducting interest the taxable income remains. The net income of the firm is the EBIT less the amount of interest and tax. Earnings applicable to the common stock (or to private investors) are the net income less dividends to the preferred stock (if there are any preferred stock which do not exist in the case of Table 6.2).

The balance sheet is a *snapshot* of the firm's assets and liabilities at the end of the year. These are usually listed in declining order of liquidity. Usually assets are current assets and long-term assets. Current assets are cash and short-term securities that can be easily liquidated, as well as receivables (which are the electricity bills that have not been paid yet), inventories of fuel, spare parts and other materials and consumables. Long-term assets consist mainly of the stock of generating plant, land, buildings, fuel storage, offices and similar facilities. The balance sheet shows the current value of these assets only in the first year. In later years this value is reduced by the amount of annual depreciation. Also, new assets can be added. Therefore future balance sheets do not show the current or actual market values of assets, or their replacement values, but rather show their depreciated value for financial statement purposes.

Liabilities of the firm are also current and long term. Current liabilities are payments that the firm will have to pay in the near future (interest payable by the firm as well as debts due in the next year). Bonds, long-term loans and similar financial instruments that will not be paid for many years are the long-term liabilities. After deducting all short- and long-terms liabilities, what is left is the shareholders' equity. This equity is equal to common stock plus retained profits. The shareholders' equity, its annual growth and its comparisons with the shareholders' paid-up capital, are important indication of the financial viability of the firm.

Most firms have to expand in the future. They can do that by self-financing, issuing new common stock or preferred stock, or by loans. Self-financing is very important for a firm's healthy growth. It is equal to retained profits and reserves plus depreciation allocations.

6.5.2   Financial ratios

Financial ratios are used to understand and test the viability of a firm, the performance and returns of its investments. In case of evaluation of individual projects, financial ratios greatly help in studying the impact of the new project on the financial and commercial outlook of the firm (see Section 6.9). There are many financial ratios of interest, the most important of which are briefly stated below.

Debt-equity ratio is a measure of the leverage of the firm

$$= \frac{\text{long-term debt} + \text{value of any leases}}{\text{equity}}.$$

In the case of Table 6.2 it equals $100/300 = 0.33$ (for the year 1998).

Times interest earned ratio demonstrates the ability of the firm to serve the interest on its debt

$$= \frac{\text{EBIT} + \text{depreciation}}{\text{interest}}.$$

Referring to Table 6.2 it equals

$$(26.0 + 25)/10 = 5.1.$$

Current ratio is a measure of the firm's potential reservoir of cash and its ability to meet its short term liabilities

$$= \frac{\text{current assets}}{\text{current liabilities}}.$$

Current ratio in Table 6.2 is $43.0/115.0 = 0.374$.

Net profit margin is the proportion of sales that finds it way into profits

$$= \frac{\text{EBIT} - \text{tax}}{\text{sales}}.$$

From Table 6.2 it equals $(26.0 - 6.4)/125 = 0.157$.

Return on total assets and return on equity are most important indicators of a firm's performance.

$$\text{Return on total assets} = \frac{\text{EBIT} - \text{tax}}{\text{average total assets}}.$$

In Table 6.2

$$= (26.0 - 6.4)/610 = 0.032$$

(average total assets equal sum of assets at the beginning and end of year divided by two and assumed to equal £610 m for year 2002).

$$\text{Return on equity} = \frac{\text{earnings available to shareholders (owners)}}{\text{average equity}};$$

this equals

$$9.6/300 = 0.032.$$

Price–earnings ratio (P/E ratio) is a much quoted measure by the financial press. It is the ratio of the price of a firm's share in the stock exchange compared with its earnings. Therefore it estimates how the firm is esteemed by investors and their expectation of its future performance.

$$\text{P/E ratio} = \frac{\text{stock (share) price}}{\text{earnings per share}}.$$

In the case of Table 6.2, if the number of shares of the firm is 300 million and the price of a share in the market is £1.05 then the earnings per share are 0.032 and the P/E = 33.

A high P/E ratio (like above) can indicate that the firm is at low risk, or there is expectation of higher growth in earnings in the future or in the value of the stock (or a combination of these).

$$\text{Dividend yield} = \frac{\text{dividend per share}}{\text{stock price}}.$$

In the case of Table 6.2 this equals

$$0.032/1.05 = 0.0305 \cong 3.1 \text{ per cent.}$$

6.5.3   Flow of the funds statement – sources and applications of funds

This important financial statement shows how the firm utilises its current assets to fund its investment and distribute dividends to its shareholders.

Internal sources of funds for a firm are operating cash flow income (after tax and interest), plus depreciation. External sources were explained before. Applications (uses) of funds are utilised for financing new investments, paying dividends as well as increase (decrease) in the networking capital.

Table 6.3 explains a Sources and Applications of Fund Statement, as applied to a firm having the following income statement.

Table 6.3   Sources and Applications (uses)
of Fund Statement

	Thousand £
Income Statement	
Income	2000
Cost of sales	1400
Depreciation	100
EBIT	500
Interest	100
Tax at 40%	160
Net income	240
Sources and Applications of Funds	
Sources	
Net income	240
Depreciation	100
Borrowing	400
Stock issue	400
Sources	1140
Applications (uses)	
New investment	1000
Dividends	100
Increase in networking capital	40
Applications (uses)	1140

Such a Sources and Applications of Funds Statement leads, with the information available from the same year's financial statement and the other market and investment data referred to above, into a *pro forma* financial statement for the next year.

Financial planning that necessitates attempts to predict the firms' financial performance involves simulation and many assumptions including prediction of sales, prices, performance of competitors, price of money, depreciation and taxation polices, etc. Such planning has to proceed year after year with the approach followed in Tables 6.2 and 6.3. A manual approach is tedious. Nowadays, standard spread-sheet programs such as LOTUS 1-2-3, EXCEL 7 and more recent versions greatly ease the work of financial planning managers.

6.6 Evaluation of benefits in the electricity supply industry

Electricity production has one output, that is electrical energy in the form of kilowatt hours (kWh). The benefits of most investments in the industry can be presented in the form of kWh. In selecting the least-cost solution projects, it is wiser to present these benefits in the form of kWh rather than money; then discount these benefits [7]. Benefits in monetary terms are kWh multiplied by the tariff, or other forms of valuation of the kWh (as in the case of reducing electricity interruption). Dealing with kWh will ease the work of the evaluator by avoiding other estimations like tariffs. This has been shown by Table 5.2. However, the following will deal with it in greater detail.

This approach is particularly suitable for evaluating base load generation alternatives. The real costs of each alternative (capital plus operation) are discounted to the base year and divided by the discounted net benefits (which are the generated kWh during the project lifetime discounted to the base year). The least-cost solution is the alternative with the least discounted real cost per discounted kWh. Such methods have the advantage of being independent from tariff evaluation and inflation.

Such an approach does not take into account system effects; the contribution of each base load variant (nuclear compared with coal, for instance) must take into consideration its system effect. Nuclear have a much higher investment cost over CCGT per kW installed. Its higher utilisation factor would save some uneconomical generation, which enhances its attractiveness. This is possible to compute through generation simulation, and assessing the impact of each alternative on the overall generation cost. This is detailed in Section 6.8. In addition, not every kWh in the electricity supply industry is measured by the same yardstick. The value of a base generation kWh is different from that of peak generation. A kWh curtailed has a much higher value than the tariff. The same applies to the network. Building a parallel transmission line will increase load transfer capability and reduce system losses, which have a value reflected in the tariff. It can greatly improve supply continuity which would reduce the amount of electrical energy curtailment, where each kWh saved will have an economical value of multiple that of the tariff.

In short, for alternative projects in the ESI, the procedure of evaluating the benefits in terms of discounted kWhs to the base year is quite a useful tool. However, it is only accurate if the system effects of each alternative are the same or if different system effects are evaluated and incorporated in the analysis.

6.7   Timing of projects

There is usually great pressure to start publicly owned projects as soon as possible. However, a project should not be started unless the evaluator is certain that the project will have a positive net present value, while utilising the opportunity cost of capital as a discount rate. Even then, it may not be the optimum time to start the project. Assessing the effect of delaying the project has to be undertaken to evaluate whether a possible net present value will be better slightly into the future. If a future net present value is possible, then the project has to be executed at the year that provides the highest net present value.

Usually, a project should be executed when the first year net benefits exceed opportunity cost of the investment, i.e. the discount rate times the project cost. If the delay causes a rise in the cost of the project, in real terms, this should be taken into consideration.

6.8   Dealing with projects with different lives and construction periods

Project alternatives may have different lives and execution periods [8,9]. The effect of the length of execution period is effectively reflected by discounting to the base year. The base year can be the project evaluation year, or the project commissioning year, with the same conclusions. For alternatives with widely different execution times (combined-cycle gas-turbine plant with typically two years for execution compared with four years for nuclear) system effects have to be considered.

There are various ways for comparing projects with different lives. If we compare a coal plant with an expected life of 40 years with a CCGT alternative with 20 years, then we repeat the investment cost of a new CCGT and add it to the cost stream at 20 years. However, things are not that simple when the expected life of the coal plant is only 30 years. In this case, we have to consider adding an investment in the form of a new fictitious plant (P_2) with a considered cost of 10 years after the end of the 20 years' life of the first CCGT plant. Since the comparison is over 30 years only, we have to calculate the cost of this new plant P_2 (which will actually live for 20 years) over only its first 10 years of service. This cost will be equal to the benefits (over the period 20–30 years) at a discount rate that is equal to the internal rate of return (IRR) of this investment. Since we know the cost of plant P_2 and can evaluate its benefits over its full life, we can calculate its IRR. This IRR is utilised as a discount rate of the benefits over the remaining 10 years of the project life (the period 20–30 years). It will be equal to cost of the fictitious plant P_2 that will (for comparison purposes) serve only over the period of 20–30 years. This evaluated cost of P_2 will be added to the cost stream after 20 years.

Discounting greatly reduces the significance of this problem, particularly if high discount rates are concerned. After 20 years, the discounted value of the repeat project is 0.149 of its present cost at 10 per cent discount rate, and only 0.061 at 15 per cent discount rate.

The annual cost method as well as the accepted method, UNIPEDE, manages the life of the alternative through calculating the discounted cost per discounted kWh. This method, also, does not reflect the technology advantage of the shorter life alternative.

6.9 Expansion projects

For a project expansion (or a firm's activity), it is necessary to carry out a *with and without* project evaluation. This involves estimating the additional benefits brought about by the project and comparing them with the additional costs caused by the project. With and without project evaluation is quite common in the electricity supply industry.

In this case, the analysis is slightly more complex. It is necessary to produce two sets of financial statements over the life of the project: one including the project expansion and another without it. The difference between the two statements indicates the incremental financial benefits brought about by the project expansion. Such incremental financial statements allow the calculation of the incremental financial IRR. After allowing for inflation, the determining factor for proceeding with the project expansion will be the incremental real IRR.

This is different from before and after comparison, because even without the project expansion the net benefit in the project area may change.

6.10 System linkages

System linkages are important in electrical power systems and were referred to more than once earlier. However, these need to be handled in more detail. The electrical power industry, more than any other industry, has the distinct peculiarity that most of its major projects have financial and economical impacts that extend beyond the confines of the project to affect the whole electrical power system. This is because of the interconnected system and synchronic performance. Any new major project, or action, will have an immediate impact that is reflected on the economics of the supply, and sometimes on its quality. Building a new modern and efficient power station would contribute to more electrical energy production and reduce the overall system costs. This is through the production reduction from costlier power stations. Halting a generating unit for maintenance can cause (beside the maintenance cost) major system costs, through the need to operate or increase the production of a less economical generation plant. Adding a new transformer or transmission line can, besides enhancing the transfer of power, reduce system losses and increase availability of supply. This, in turn, will reduce the economic and social cost to the consumers of electricity interruptions.

Therefore, in all major power projects, particularly generation, it is advisable to perform a computer simulation over the next few years, with and without the project. The simulation output of the project is integrated in the system performance. This will allow a better evaluation of the financial and economical impacts of the project. Details of the system simulation are fully explained in published literature [10] and will also be covered in Chapter 10. A flow chart of generation costing and expansion simulation is depicted in Figure 6.1.

Many projects in the electricity supply industry are undertaken to reduce supply interruption and improve service to consumers. In these instances, it is necessary for

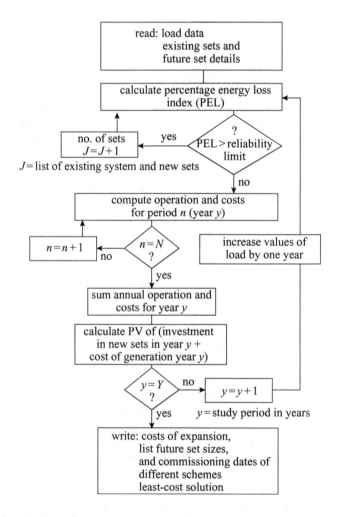

Figure 6.1   Costing of a system expansion plan

the electrical utility to put a monetary value on each kWh curtailed. The economics of the network strengthening projects are evaluated through comparing the discounted annual value of the reduction in electricity curtailment (annual kWh saved multiplied by the social cost of each kWh interrupted) with the annual cost. This will be dealt with in detail in Chapter 10.

6.11 References

1 CHRISTIAN D. J., and PRASAD, R.: 'Guidelines for calculating financial and economic rates of return for DFC projects' (International Bank for Reconstruction and Development, Washington DC, 1984)
2 SQUIRE, L., and VAN DER TAK, H.: 'Economic analysis projects' (The Johns Hopkins University Press, 1989)
3 FELDER, F.: 'Integrating financial theory and methods in electricity resource planning', *Energy Policy*, February 1996, **44**, (2)
4 RAY, A.: 'Cost benefit analysis' (The Johns Hopkins University Press, 1990)
5 BREALY, R., and MYERS, S.: 'Principles of corporate finance' (McGraw-Hill Inc, 2002)
6 BRIGHAME, E.: 'Fundamentals of financial management' (The Dryden Press, 1989, 5th edn.)
7 'Projects and costs of generating electricity' (OECD, Paris, 1989)
8 MORRIS, T. E.: 'Economic comparisons in planning for electricity supply', *Proc. IEE*, March 1970, **117**, (3)
9 DAVISE, B. G. (Ed.): 'The economic evaluation of projects' (The World Bank, 1996)
10 KHATIB, H.: 'Economics of reliability in electrical power systems' (Technicopy, Glos, 1978)

Chapter 7

Economic evaluation of projects

7.1 Introduction

The financial evaluation of a project studies its performance and return (profitability) from the points of view of the industry (the utility and the firm), the owner and the investor. Economic evaluation, on the other hand, studies project benefits and returns from the national economy point of view and assesses the effect the project will have on the overall economy of the country. While small projects have limited, if any, impact on the national economy, large projects have social and environmental effects, which cannot be ignored (referred to earlier as externalities). The electrical power industry is highly capital intensive. A modern power station normally costs hundreds of millions of pounds, and so do some network projects. In high-capital projects of this nature it is necessary to find the least-cost solution not only from the point of view of the industry but from that of the national economy as well. The project evaluation should not be restricted to its profitability and should also evaluate its national economic and environmental impacts (social impacts) and compare them with those of other competing projects in the resource-limited national economy to ensure economic efficiency.

Economical analysis of major projects, therefore, evaluates two things:

(1)   the priority of the project in the national plans of the country and
(2)   its effect on the overall economy of the country.

The ultimate purpose of the analysis would be to provide a measure of the impact of the project on the national welfare as exemplified by increased consumption of goods and services (including electrical energy) that serve as a proxy to increased welfare.

Economic efficiency is important to society and has four primary components. The first component is the requirement of production efficiency, or incurring the minimum feasible cost in producing goods and services sold. Second, efficient variety is needed, meaning that the menu of goods and services offered is tailored to the wants and needs of customers. Third, society wants allocative efficiency, where the goods and services go to customers who value these goods and services most highly. Finally, society wants dynamic efficiency, which means that the first three types of efficiency are sustained [1].

In the electricity supply industry the first component is fulfilled through the least-cost solution alternative. The second and third purposes are achieved through the implementation of appropriate electricity tariff and security levels to different classes of consumers. The last purpose of sustainability is attained through integrated resource planning.

Economic efficiency (which takes into consideration social as well as environmental costs and benefits) should be a fundamental criterion of public investment and policy making. It should also be fundamental for large capital-intensive private sector investment that replaces, supplements or competes with public investment. This includes investments made by independent power producers. If real costs and benefits of large projects are not addressed from the perspectives both of the industry and of the whole economy the wrong projects (or alternatives) may be selected.

The same method of discounting costs and benefits is used for both financial and economic evaluation. Economic evaluation is concerned with estimating the economic rate of return (ERR). Financial analysis, on the other hand, evaluates the internal rate of return (IRR) and considers only financial parameters in the analysis. A project, for example building a power station firing cheap low-quality coal and located near the load centre, may have a high financial IRR. But when environmental costs of the project are considered they may result in low (or negative) ERR and may lead to the rejection of the project on economic rather than financial grounds [2].

In many cases market prices do not reflect the real cost of resources, products or services to the economy. Market prices are usually distorted by duties, taxes, subsidies and other trade restrictions. Such instruments have a major influence on the financial profitability of a project from the industry or investor's point of view but none on its economic viability from the perspective of the national economy. Distortions, particularly subsidies, vary from one country to another but are more prominent in the developing countries than in market economies where efficient and transparent prices prevail. They do exist in every country in varying degrees, particularly in energy fuels and products prices. This is most vividly exemplified in the electrical power industry where in most countries prices of fuels are set at a level far from their true cost to the economy.

A lot of work has already been done during the past three decades to establish the techniques for economic evaluation of projects. Most of this work was done by the international lending agencies like the World Bank, OECD, ODA and various United Nations agencies [3–7]. Handling detailed mechanics of economic evaluation requires a skilled economist. These are briefly explained below. An important issue, which project planners need to be aware of, is the existence of distortions and the general techniques of dealing with them. Four issues are important: transfer payments, border and shadow pricing, externalities, and system linkages. Two other related issues are also important: integrated resource planning (this was dealt with briefly in Chapter 1) and environmental issues. The environmental impact of the electricity supply industry is significant. This is an externality that needs thorough evaluation and understanding and is dealt with separately in Chapters 8 and 9.

In most of the analysis that follows it is assumed that the official exchange rate of currency is not artificially fixed (overvalued) but rather is freely adjustable according to international trading links. This is the case in all OECD countries and is becoming increasingly so in most developing countries.

7.2   From financial evaluation to economic evaluation

The two time streams of financial costs and benefits form the starting point for economic evaluation. These financial streams have to be adjusted and modified with regard to the concepts of transfer payments, border and shadow pricing, and externalities. The essence of sound economic evaluation is to remove distortions, particularly price distortions, to improve investment decisions. Such distortions are mainly caused by transfer payments (subsidies, taxes and levies), imperfections in the domestic market (monopolies, restrictive practices, etc.) and distortions caused by trade policies (exchange rates, quotas, etc.).

To shift from financial to economic evaluation it is essential to implement a number of steps:

(a)   remove direct transfer payments,
(b)   use border prices in the case of traded goods and services (particularly fuels),
(c)   use shadow prices, particularly in dealing with non-traded items (such as land, unskilled labour, etc.),
(d)   account for externalities (especially environmental impacts and costs), and
(e)   allow for the difference between accounting and the official price of foreign exchange, if any.

Therefore, in economic evaluation, transfer payments (taxes, duties, subsidies) are not included. Costs and benefits are priced in border (international) prices for traded goods and shadow prices for non-traded goods. Externalities, whenever they exist, are added to costs and benefits.

7.3   Transfer payments

A transfer payment is a payment made without receiving any good or direct service in return. In economic evaluation of projects the most-encountered transfer payments are taxes and direct subsidies, as well as loans and debt service (payment of interest and repayment of principal). These are treated as transfer payments because the loan terms only divide the claims to goods and services between borrowers and lenders and do not affect the total amount of true return to investment. Transfer payments represent only shifts in claims to goods and services and not use of new production. They do not increase or reduce national income and are hence omitted when converting financial streams used in financial evaluation to economic values used in economic evaluation [8].

In the electricity supply industry, transfer payments normally apply to local duties, fuel subsidies (or taxes) and other subsidies, taxes and interest. Such payments do not represent direct claims on the country's resources but merely reflect a transfer of control over resource allocations from one part or sector of the national economy to another, and should therefore be omitted from economic evaluation. In contrast, they are quite important in financial analysis.

Economic analysis is not concerned with sources of investment funds (equity or loans or a combination of these) and repayment of loans (particularly local loans), since similarly the loan and its repayment are financial transfers and are not part of the economic analysis. The economic cost of a project is its investment cost in the base year less its discounted terminal value if any, converted into economic terms as detailed below. It is, however, essential to distinguish between taxes that are transfer payments and are ignored in economic analysis, and other taxes that are a payment for goods and services and should be included. For instance, a road tax can be a payment for the services provided by the road, and a municipality tax can involve a charge for sewerage and should therefore be included in the analysis.

Subsidies are direct transfer payments that flow in the opposite direction from taxes. Fuel subsidies are common in the electricity supply industry, particularly for local coal supplies. Conversely, imported sources of fuel may be heavily taxed to encourage local energy sources, possibly for environmental reasons, or merely to provide an income for the government. In such cases local energy sources have to be priced at their full actual cost to the economy. If these items are tradable (as will be explained below) then their set prices should reflect border prices if they are higher than their production cost.

The same logic applies to credit transactions – loans and loan servicing. Such transactions do not increase national income but again represent the transfer of control of resources from one sector of the economy to another, i.e. they are transfer payments. Foreign loans that impose a burden on foreign reserves or that are subsidised by an undervalued currency have to be considered in the economic evaluation.

7.4   Externalities

An externality is an effect of a project felt outside its confines and is not included in its financial evaluation. Externalities may be either technological or pecuniary. An example of a technological externality might be pollution from a power station that causes direct material and health damage. Economic evaluation usually tries to incorporate technological externalities, especially costs, within the project account and thus change them from externalities to project costs and benefits (internalise them). For example, costs of the damage to buildings and health by sulphur dioxide (SO_2) emissions, and to water basins caused by the waste discharges from the power station, may be calculated and assigned to the project. Pecuniary externalities arise when the project affects the prices paid or received by others outside the project, for example, where building a new efficient and clean power station reduces prices of electricity to users. Pecuniary externalities have to be included in both project financial and economic analysis.

Externalities are not easy to define and are very difficult to quantify. Their identification and attempt at quantification is an important part of the job of a skilled project evaluator. Some externalities are positive. In the electricity supply industry they may be in the form of technology advancement, export promotion, job creation, training, etc. Examples of negative externalities include detrimental environmental impacts and congestion.

Most large projects in the electrical power industry have externalities, particularly environmental impacts. These range from minor visual impact and landscaping problems in small transmission and distribution projects to very serious pollution problems, as in low-quality coal power stations. In most cases such environmental aspects are identified in an impact assessment study and attempts are made to rectify them in project design and costing. However, other intangible effects, like visual impact, noise level, congestion, etc., are very difficult to quantify, as are some of the benefits like technological development, technology transfer and human capacity building. Still they should not be ignored and every effort should be made to identify and quantify them whenever possible. Externalities can have a strong influence on the choice of the least-cost solution, particularly when the difference in cost between alternatives is small.

It is important to try to identify all externalities caused by the project. Where there are significant externalities an attempt should be made to quantify them and include their values in the project costs and benefits. Where it is difficult to quantify them, as usually is the case, they should be cited in the project economical evaluation and they may affect the choice of the least-cost solution. Environmental externalities of electricity production are particularly important and these are dealt with in detail in the following chapters.

A main externality aspect of a project in the electricity supply industry is the system linkage of projects. System linkages are external to the project itself but are internal to the electricity supply industry as a whole, and therefore cannot be treated as a true externality. They should be taken into consideration in all financial and economic evaluation of projects.

7.5　Border and shadow pricing

During construction each project incorporates a lot of inputs (resources) such as equipment, materials, land and manpower. During the lifetime of the project it also consumes inputs, and produces output (product). Among the most important operational inputs in power projects are fuel, manpower and similar resources. In the financial evaluation these inputs are valued at their market prices. Market prices are almost never ideal and are often distorted to varying degrees in different countries. The industry and investors are mainly interested in market prices of inputs and outputs because that is what they have to deal with and it is these prices that determine financial profitability. Economic evaluation goes beyond this by investigating the true impact of the project on the national welfare. To achieve this, prices of resources must be set at their true cost to the economy [9].

Inputs and outputs can be classified into two categories: traded and non-traded. A project input or output is deemed to be traded if its production or consumption will affect a country's level of imports or exports at the margin. A project input or output is considered to be non-traded because of its bulkiness, cost consideration and immobility or other restrictive trade practices. Machinery and equipment, as well as fuel and skilled labour and marketable products, are considered tradable. Non-traded goods and services include items like land, water, buildings, unskilled labour, electricity (in most cases) and many other services and bulky material. Traded and non-traded goods and services are part of project inputs and outputs. For economic evaluation they have to be priced at their true cost to the economy.

7.5.1   Pricing traded inputs and outputs

Because they can be traded, i.e. imported as well as exported, tradable inputs and outputs create a change in the country's net import or net export position at the margin. They must be valued, in economic evaluation, at *border prices*. Border prices are world prices, *free on board* (*FOB*) for exports and *cost, insurance and freight* (*CIF*) landed cost for imports, adjusted by allowing for domestic transfer costs. Transfer costs are costs that are incurred in moving inputs and outputs between project site, border and target markets.

Take coal as an example. The border price of imported coal, which will be incorporated in the economic evaluation, will be the CIF price at the nearest port plus handling and transport charges to the generating plant. If this coal is produced locally, its economic price will be its FOB price at the port of export minus the cost of transport from the coalmine to the port plus the cost of transport to the generating plant. For bulky materials, like coal, such transport prices are significant and greatly favour local production.

Consider an example in which coal can be imported by country 'y' at $80 ton^{-1} with transport cost between country 'x' and country 'y' at $20 ton^{-1}. If country 'x' would export its production to country 'y' then it has to price it at $60 ton^{-1} ($80 − $20 ton^{-1}). If the cost of handling this coal and transporting it to the export port of country 'x' is $15 ton^{-1}, then coal at the mine-mouth has to be $45 ton^{-1}. If this coal will be used locally instead and with transport plus handling cost of $5 ton^{-1} from the local mine to the local power station then the economic cost of coal, will be $45 + $5, i.e. $50 ton^{-1}, irrespective of the actual cost of extraction, which can be much less.

If the cost of production of coal in country 'x' increases and it considers importing outside coal from a source with a sea transport cost of $15 ton^{-1}, then its cost CIF of imported coal will be $50 + $15 = $65 ton^{-1}. If transport of this coal to the power station will involve another $8 ton^{-1} then the total cost of imported coal to the power station will be $73 ton^{-1}.

Country 'x' is advised to continue production of coal as long as its production cost is equal to $68 ($73 − $5 ton^{-1} local transport cost from mine to power station). If country 'x' decides to continue producing coal from its mines, even if its production cost reaches $90 ton^{-1}, while selling it at $70 ton^{-1} to the power station, then its coal subsidies will amount to $20 ton^{-1}. Such subsidies are excluded in the economic evaluation. The market price for coal will be $70 ton^{-1} in the financial evaluation of the investors, and $73 ton^{-1} (i.e. the border price) in the case of economic evaluation.

A decision to utilise a local resource (like coal) is not dependent only on production costs and border prices but is also influenced by many other political and social considerations, like utilising local resources, creating local employment, supply security and shortage of foreign exchange. In making such decisions, however, the border price of coal has to be calculated and taken into consideration in evaluating plans and decisions.

7.5.2   *Pricing non-traded inputs and services*

Non-traded goods and services are priced at their *shadow price*. A lot of research was done in understanding and evaluating shadow pricing [3–6]. With the enhancement of free trade, freely convertible currencies and open markets, shadow pricing is still required but not to the extent it was in the past.

In evaluating non-traded goods and services it is essential to differentiate between non-traded tradables and non-tradables. Non-traded tradables are goods and services that can be traded on the international market but are not because of their cost being higher than international prices (e.g. local low quality coal), or because of trade restrictions and policies (quotas, restrictive import taxes at potential import markets, etc.).

A good example of a non-traded tradable is coal. If the international price of coal is $50 and its transport to the power station involves an additional $23, a local coalmine with a high production cost of $68 (which is above international prices hence rendering its product a non-traded tradable) will continue production protected by the high transport prices. In such cases the economic price of a non-traded tradable commodity is the opportunity cost of the product, that is the price it can command in the absence of the project.

Most power projects involve non-tradable inputs, mainly materials for civil works and labour. Such inputs can be decomposed into its components. Large civil works like a power station building contain tradable and non-tradable components. Tradable inputs (cement, steel, etc.) can be priced in accordance with their border prices. Non-tradable inputs (gravel, sand, stones, labour, etc.) have to be shadow priced through utilising conversion factors [9]. The most widely used factor is the standard conversion factor (SCF). This factor is the average ratio of border and domestic market prices and is equal to

$$\frac{M + X}{(M + T_m) + (X - T_x)}$$

where M = CIF value of imports, X = FOB price for exports, T_m = all taxes on imports, T_x = all taxes on exports. Through applying this conversion factor by multiplying it by the non-tradable inputs it is possible to reduce the impact of local distortions. In a way the SCF is the ratio between an official and a shadow exchange rate. In developing countries it is usually less than unity, signifying that the local currency is overvalued.

Consider a country with imports valued at 100 million currency units and exports of 50 million units, import taxes of 20 million units and export taxes of zero. The SCF is equal to 0.88 (which is $(100 + 50)$ million divided by $[(100 + 20) + (50 - 0)]$ million). It is also usual to have a SCF equal to 1.0 for skilled labour and 0.5 for unskilled labour.

An example is a power station building, which is priced in local currency as shown in Table 7.1. The conversion factors for steel and cement were calculated using border prices as shown. The economic cost of the power station building will be as detailed in local currency. An estimation of economic costs and benefits is given in Table 7.2.

Table 7.1   Conversion of financial cost of a power station civil work component into economic cost (in local currency units)

Component	Financial cost	Conversion factor	Economic cost
Traded items			
Cement	10 000	0.90	9000
Steel	30 000	0.80	24 000
Non-traded items			
Other building materials	10 000	0.88	8800
Overhead	20 000	0.88	17 600
Unskilled labour	25 000	0.50	12 500
Skilled labour	5000	1.00	5000
Total	100 000	0.77	76 900

Table 7.2   Estimation of economic costs and benefits [9]

Project inputs
    A   Net imports of tradable items (CIF plus converted port-to-project costs)
    B   Diverted net exports of tradable items (FOB minus converted source-to-port plus converted source-to-project costs)
    C   Non-traded items
        1   Land (converted opportunity cost)
        2   Labour
            (a) Skilled (market wage rate)
            (b) Unskilled (converted shadow wage rate)
        3   Goods (converted domestic market price plus converted source-to-project costs)
Project outputs
    A   Net exports (FOB minus converted project-to-port costs)
    B   Import substitutes (CIF plus converted port-to-market costs and minus converted project-to-market costs)
    C   Non-traded items (converted factory-gate price)

7.6   References

1   FEDLER, F.: 'Integrating financial theory and methods in electricity resource planning', *Energy Policy*, February 1996, **24**, (2)
2   KHATIB, H.: 'Financial and economic evaluation of projects', *Power Eng. J.*, February 1996, **10**, (1)
3   SQUIRE, L., and VAN DER TAK, H.: 'Economic analysis of projects' (The Johns Hopkins University Press, 1989)
4   RAY, A.: 'Cost-benefit analysis' (The Johns Hopkins University Press, 1990)
5   LAL, D.: 'Methods of project analysis' (International Bank for Reconstruction and Development, Washington DC, 1976)
6   LITTLE, I. M. D., and MIRRLFES, J. A.: 'Project appraisal and planning for developing countries' (Heinemann Educational Books, 1974)
7   'Appraisal of projects in developing countries' ODA (Overseas Development Administration), (HMSO, 1988, 3rd edn.)
8   GITTNER, J. P.: 'Economic analysis of agricultural projects' (The Johns Hopkins University Press, 1982, 2nd edn.)
9   CHRISTIAN, D. J., and PRASAD, R.: 'Guidelines for calculating financial and economic rates of return for DFC projects' (International Bank for Reconstruction and Development, Washington DC, 1984)

Chapter 8

Environmental considerations and cost estimation in project evaluation

8.1 Introduction

Electricity utilisation is environmentally benign and as a form of energy carrier electricity is clean and safe. It causes no pollution or environmental emissions at the point of use. It was also proven that electricity can be more efficient than other forms of energy [1]. Therefore substituting electricity for other forms of energy can greatly help in reducing global emissions and pollution caused by the use of the latter. In addition, the fact that electricity production is undertaken at a single point, namely the power station site, means that environmental problems associated with electricity production are concentrated at a single point, which makes containing and dealing with them much easier.

Electricity production can cause local and regional environmental impact and may also have long-lasting detrimental global consequences. Some of these impacts like the emissions of sulphur dioxide (SO_2), nitrogen oxides (NO_x) and solid particulates, which all have detrimental air quality implications, can be controlled by investing in technology and abatement facilities. These measures can control and reduce significantly such emissions. Carbon dioxide (CO_2), which is the main gas suspected of causing global warming (greenhouse effect), is far more difficult and expensive to control. More than half of thermal generation and 39 per cent of total electricity production uses coal (sometimes low-quality coal and lignite). Coal use for this purpose is increasing particularly in developing countries [2]. These countries, with their limited financial resources, cannot afford to invest in expensive measures for environmental preservation, particularly those concerned with long-term global effects. Their main concern is the production of the largest possible quantity of electricity to meet growing local demand in the most extensive manner and as cheaply as possible. Their environmental actions, in most cases, are a response to an impending major environmental disaster [3].

Employment of abatement technologies is not the only way to control emissions. Substantial emission containment can be achieved through enhancing efficiency in production, conservation of use, the use of cleaner fuels (natural gas) and clean technologies (as detailed later), and through the use of non-fossil energy sources (hydro, nuclear and renewables). The major effort that has taken place recently in the field of improving efficiency of electricity generation was discussed in Chapter 1. Developments in this field are still taking shape and will ultimately lead to the production of more electricity with less quantity of fuel. Gradual improvements in the energy sector are taking place where electricity is substituting other forms of energy in final use. This will enhance the role of electricity in environmental preservation by reducing emissions from other sources. Clean coal technologies are still being developed [4]; carbon sequestration technologies, although showing promise, are still in their infancy and it will be many years before they become effective.

Conservation in consumption as well as improved efficiency in electricity utilisation are important means for attaining the environmental goals. Conservation and efficiency-in-use measures do not always require large investments and should therefore be the first technologies and methods to be applied. But they do require policies. The main promoter of these measures is the pricing mechanism (tariffs). Another powerful way of promotion is through informing and educating consumers and soliciting their participation in conservation and efficiency measures and in investing in energy-saving technologies. Heat pumps, electric arc melting, induction heating, electric transit, and efficient electric house appliances and modern electricity-saving lighting can significantly reduce energy and electricity consumption without compromising consumers' production and welfare. Such electricity conservation, and efficiency in use, plus switching from primary energy use into electricity, lead to substantial reductions in emissions and environmental impacts of electricity and energy without having to invest in capital-intensive environmental technologies.

Because of the nature of fuels utilised in the electricity supply industry and the rapidly growing utilisation of energy in the form of electricity, the environmental impacts of electricity production are gaining more importance than those of other forms of energy and this demands a thorough understanding of their nature. Such impacts can have direct financial costs (damage to trees and forests and pollution to land and other resources that can be financially assessed) as well as intangibles (mainly health). These damages can be grouped as social costs. There are also the possible long-term environmental costs of global warming. Some of these environmental impacts have to be avoided through investment. A cost/benefit analysis to the costs involved in controlling emissions from the electricity supply industry and the benefits derived out of this action have to be undertaken. Unfortunately, a lot of environmental damage costs resulting from electricity production are very difficult to quantify and the science of environmental accounting is still in its infancy [5]. This chapter will try to record the environmental detrimental effects of electricity, methods of quantifying them (where possible) and analyse technologies, investments and other means of control.

8.2 Environmental impacts of electricity fuels

Electricity production can have many environmental health and safety impacts. Mining for electricity fuels, their transport and storage also contribute to pollution. So does the disposal of the combustion-process products of ash and other solid wastes. Coal, in environmental terms, represents particular concerns. Its consumption by-products are more polluting than other fuels and they contain higher carbon, SO_2 and NO_x emissions and much more solid waste. Environmental health and other safety impacts of coal mining and transport can be substantial.

The main categories of impact of pollution are local, regional and global.

(i) Local impacts are mainly in the form of heavy hydrocarbons and particulate matter (including sulphur flakes), which are deposited within hours and can travel up to 100 km from the source.

(ii) Regional impacts involve emissions and effluents, the most important of which are SO_2 acid depositions, which have a residence time in the atmosphere of a few days and may travel to a few thousand kilometres thus causing cross-boundary effects.

(iii) Global pollution is exemplified by CO_2 emissions and other gases (mainly methane), which have long residence times in the atmosphere. Methods of costing the three impacts are described in the following section. The various impacts of the fuel system components are detailed in Table 8.1. Carbon emissions of the fossil fuels are shown in Table 8.2, and the global warming potentials (GWP) of different emissions are in Table 8.3.

Not all impacts of electricity utilisation are detrimental. Substituting electricity for other forms of energy helps in improving the efficiency of use of primary fuels. Electricity has replaced most other fuels in urban societies and has improved the quality of urban and built-up environments and of environmental management through various instrumentation and control technologies. The fact that electricity is highly controllable significantly improves the efficiency of heating and cooling systems, as well as machinery and lighting through shorter warm-up and cool-down periods. Therefore electricity can play a significant positive role in safeguarding local, regional and global environment if measures to control the detrimental supply side impacts are dealt with in an economically sound manner.

Global warming potentials (GWP) show that greenhouse gas emissions (GHGs) other than CO_2 can pose a much larger risk to the environment if emissions continue unchecked. GWPs are a measure of the relative radiative effect of a given substance compared to CO_2, over a chosen time horizon. They are an index for estimating the relative global warming contribution due to atmospheric emission of 1 kg of a particular GHG compared with 1 kg of CO_2. The atmospheric response time of CO_2 is subject to substantial scientific uncertainties, owing to limitations in the understanding of key processes, including its uptake by the biosphere and ocean. The numerical values of the GWPs of all GHGs can change.

Table 8.1   Environmental impacts of fuel system components

Electricity generation fuel	Key impact
Coal	Groundwater contamination
	Land disturbance, changes in land use and long-term ecosystem destruction
	Emissions of SO_2, NO_x, particulates with air quality implications, heavy metals leachable from ash and slag wastes
	Possible global climatic change from CO_2 emissions
	Lake acidification and loss of communities due to acid depositions
Oil and gas	Marine and coastal pollution (from spills)
	Damage to structures, soil changes, forest degradation, lake acidification from S and N emissions
	Groundwater contamination
	Greenhouse gas emissions impact, e.g. possible global climate change
Hydroelectric	Land destruction, change in land use, modification of sedimentation
	Ecosystem destruction and loss of species diversity
	Changes in water quality and marine life
	Population displacement
Nuclear	Surface and groundwater pollution (mining)
	Changes in land use and ecosystem destruction
	Potential land and marine contamination with radionuclides (accident conditions)
Renewable	Atmospheric and water contamination
	Changes in land use and ecosystem
	Noise from wind turbine operations
	Possible air quality implications

Source: Reference [6].

Table 8.2   Carbon emissions of fossil fuels

	Tonnes of carbon per TOE
Coal	1.08
Fuel oil	0.84
Natural gas	0.64

TOE = tonnes of oil equivalent.

Source: Reference [7].

Table 8.3   Global warming potentials (GWP)

Gas	Global warming potential time horizon		
	20 years	100 years	500 years
Carbon dioxide (CO_2)	1	1	1
Methane (CH_4)	62 (56)	23 (21)	7 (6.5)
Nitrous oxide (N_2O)	275 (280)	296 (310)	156 (170)
Sulphur hexafluoride (SF_6)	15 100 (16 300)	22 200 (23 900)	32 400 (34 900)

Source: IPCC (2001) (1995 IPCC figures in brackets).

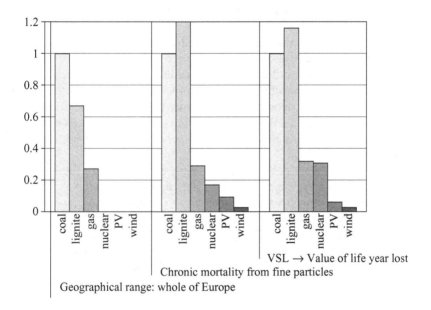

Figure 8.1   Relative ranking of external cost estimates (excluding global warming impacts) for different electricity generation technologies under changing background assumptions. External costs are normalised to the external costs for the coal fuel cycle [8]

8.2.1   Environmental impact of different generation fuels

The ExternE project is a major research programme launched by the European Commission at the beginning of the 1990s to provide a scientific basis for the quantification of energy related externalities and to give guidance supporting the design of internalisation measures [8].

In spite of the existing uncertainties, ExternE results were quite robust with regard to the relative ranking of different electricity generation technologies. Figure 8.1

clearly shows that even under different background assumptions electricity generation from solid fossil fuels is consistently associated with the highest external cost, while the renewable energy sources cause the lowest externalities.

8.3 Environmental evaluation

It is not easy to estimate the cost of environmental impacts of electricity production, local or regional. Local environmental impacts depend on the quality of fuel utilised, the site of the power station with regard to population centres, and the availability of other amenities. A power station sited in the desert will have much less impact on environmental health than a similar power station close to population centres and forests. A hydro-electric power station in an uninhibited area will have a lot less environmental impact than a hydro-power station that leads to relocating people or to the destruction of villages and agricultural lands.

Environmental impacts that can be quantified in financial terms should be incorporated in the economic evaluation of the project. Those that cannot be directly evaluated (such as loss of bio-diversity, impact on health, etc.) should be researched and quantified, whenever possible, and also incorporated as social costs. Such costs have to be utilised in the economical evaluation used to identify the least cost and in the calculation of the IRR and ERR. Detrimental environmental impacts can be lessened or alleviated through investment and a cost–benefit analysis of such investments should be performed.

Today, emissions from power stations have been identified and most of them can be evaluated in terms of their environmental impacts (see Table 8.1). Therefore it is now possible, although somewhat difficult, to determine the cost of a power facility's environmental impact. This cost has to be compared with the capital investment needed for implementing suitable abatement technologies. Benefits of capital investments incorporating different abatement technologies – and thus leading to different impacts – should be compared to help identify the least (economic) cost project alternative.

8.4 Environmental impacts and the discount rate

The utilisation of the discount rate in evaluating environmental impacts was subject to severe criticism on the grounds that environmental damage and costs appear many years after project execution. As long-term costs they are reduced to insignificance by discounting (at a discount rate of 10 per cent, damage valued at £1 now and occurring after 10 years will equal only £0.39 today). At the same time, project investments that help the environment will also have long-term benefits, which are greatly reduced by discounting. High discount rates also lead to accelerating the exploitation of renewable and non-renewable natural resources, such as fossil fuels, because they substantially reduce their long-term value [9].

One strong reason favouring the use of the present value of environmental costs and benefits from different perspectives over the use of the discount rates for investment by individuals is that individuals are mortal while societies are quasi-immortal. Therefore individuals, being mortals, tend to value the present in greater terms than societies, because of the prospect of their early death. Such fears do not exist in the community. The community, therefore, has reason to discount the future to a lesser extent than individuals [10].

This argument calls for lowering the discount rate on environmental projects or even not discounting their costs and benefits at all. This, of course, can be a source of difficulty in evaluations where it will result in having more than one discount rate for different cost/benefit items and implies a different treatment of environmental projects. Another approach would be to properly value, in economic and not just financial terms, environmental costs and benefits and give them the appropriate weight in project evaluation. Projects should be designed not to cause serious environmental damage to irreplaceable critical natural capital. Any long-term change in the expected relative value of environmental assets should also be reflected in their appraised prices.

Therefore cost–benefit analysis of environmental projects is better made at the same discount rate as that used for standard investment evaluation. However, more thorough research and estimation have to be undertaken of all environmental costs and benefits, particularly those of non-monetary and irreversible consequences, and the values incorporated into the analysis. Having said that, climate change is a long-term problem, so that from an economist's perspective discounting is a very important issue. In such a case three different discount rates (0, 1 and 3 per cent) are normally utilised. Although the models cover a time horizon of 100 years only, the change in discount rate – depending on the valuation alternative – might even lead to a change in the sign of the aggregated marginal damage costs (short-term benefits might dominate the global estimate when using the 3 per cent discount rate, with damage occurring after 100 years having a 5 per cent value today) [8].

8.5 Health and environmental effects of electricity

The vast majority of the environmental and health effects of electricity are from the generation side, particularly when firing low-quality coal. Hydro-electric and natural-gas fired facilities have lower environmental impacts. Applying adequate safeguards have rendered nuclear power stations safe. Sections 8.5.1 and 8.5.2 briefly detail the health and environmental impacts of electricity generation facilities.

8.5.1 Direct health effects

Fossil-fired plants

In the case of oil- and coal-fired plants, a significant public health risk results from exposure to the large amounts of gaseous and solid wastes discharged in the combustion process. These emissions include SO_2, CO, NO_x, hydrocarbons, and polycyclic organic matter. Coal-fired stations also discharge fly ash, trace metals, and radionuclides. The presence of these pollutants leads to increased incidence of respiratory disease, toxicity and cancer. Disposal of the resulting solid waste leads to health risks associated with leachate and groundwater contamination. Natural gas-fired plants pose a public health risk of NO_x and particulate emissions, but are significantly less hazardous to health than oil- or coal-fired plants.

Renewable

Large-scale hydro-electric plants pose relatively few health risks compared with fossil-fired plants. However, still water reservoirs create ecological environments favourable to the spreading of parasitic and waterborne disease. Relatively low pub lic risks exist from the low probability of dam failures. Biomass plants emit lower levels of SO_2 compared with oil- or coal-fired plants, but have higher emissions of potentially carcinogenic particulates and hydrocarbons. Other renewables such as solar photovoltaic and thermal, geothermal, tidal, and wind power, pose no significant occupational or health risks at the generation stage.

Transmission and distribution

The health impacts of transmission and distribution, particularly ultra high voltage alternating current (UHVAC) and high voltage direct current (HVDC), are well documented [11]. It has been verified that human sensitivity to short-term exposure to a strong electrostatic field could be significant. Studies to measure the biological effects of electric fields and of chronic exposure to electric fields on people, animals and plants have been carried out. Since 1981 dozens of studies examined the incidence of cancer among workers exposed to electric and magnetic fields (EMF). Results were inconsistent. While some studies showed a slight rise in mortality of leukaemia or brain cancer among electrical workers others showed no such tendency. Recent more extensive and detailed studies and research were also inconclusive. Any negative health effects of transmission and distribution may be relatively minor compared to the health effects of the alternative option: siting power generation facilities closer to population centres [12].

8.5.2   Environmental impact

As mentioned earlier, pollution resulting from electricity generation can be broadly categorised as having local, regional, and global environmental consequences. Regional and global impacts are caused primarily by the emission of atmospheric pollutants that have longer residence times, causing dispersal over larger areas. Most important among these gases are SO_2, which causes acid deposition or 'acid rain', and CO_2, which is a 'greenhouse gas' and can contribute to global warming.

Local impacts

Purely local impacts include those caused by fossil-fired power plant emissions to the atmosphere (particulates, leaded compounds, volatile organic compounds, dust) that result in air quality degradation causing damage to crops, structures, local ecosystems and posing a health hazard. This is also the case with emissions from waste-to-energy plants. Effluent disposal from fossil-fired and nuclear plants can lead to groundwater contamination with long-term irreversible pollution implications. The existence of potential carcinogens and mutagens in the waste can have negative impacts on health and agricultural productivity. Improper disposal of radioactive waste from nuclear plants (e.g. discharge of liquid waste into the sea) can cause destruction of fisheries and other health hazards through contamination of the water supply.

Hydro-electric power generation is perhaps the one electricity generation system that has only local, but occasionally major, environmental consequences. These consist of the damage caused by dam construction: destruction of habitats and loss of local/national biodiversity, the inundation of productive land and forests, and possibly the loss of cultural sites and mineral resources. Watershed disturbance sometimes leads to increased flooding and low flow in the dry season. On major river systems, this can have inter-regional and/or transitional consequences causing significant political and social unrest over water rights. The existence of still water contributes to the spread of waterborne and parasitic disease. The massive displacement of people that is often required represents a significant social cost and can lead to increased use of marginal lands.

Environmental impacts from other renewables are related primarily to the loss of land use represented by the high space intensity of solar energy, and the noise caused by wind-powered generation.

Regional impacts

Coal- and oil-fired power stations emit significant amounts of sulphur dioxide and nitrogen oxides to the atmosphere. The transport of sulphur dioxide occurs over long distances (greater than 1000 km), causing the deposition of emission products over national boundaries. This may result in ecologically sensitive ecosystems receiving depositions of sulphur well above carrying capacity.

Acid deposition caused by sulphur and nitrogen oxides results in damage to trees and crops, and sometimes extends to acidification of streams and lakes, resulting in destruction of aquatic ecosystems. It also leads to the corrosion, erosion, and discolouration of buildings, monuments and bridges. Indirect health effects are caused by the mobilisation of heavy metals in acidified water and soil.

While electricity generation accounts for less than half of the total anthropogenic nitrogen oxide emissions (the majority of the remainder is caused by motor vehicle exhausts), the portion of SO_2 emitted by electricity generation is substantial. For example, Europe (including Eastern Europe) emitted close to a third of the worldwide anthropogenic sulphur emissions in the early 1980s. It is estimated that in 1985, 60 per cent of European SO_2 emissions were the result of electricity production.

Other regional environmental impacts are caused by radiation effects on health, and land/water contamination caused by severe nuclear accidents. As mentioned before, changes in hydrological flow and water conditions caused by dams can also have regional consequences. Thermal-power plants could also have a negative effect on aquatic organisms, fisheries, etc., owing to the water temperature increase that could be caused by diffused thermal effluent.

The relative contribution of electricity generation to the prospects of overall global warming (mainly in the form of CO_2 emissions) has been estimated at about 20 per cent till now, but rapidly increasing, compared with about 7–8 per cent caused by deforestation. Of the contribution of fossil fuels, coal and oil each contribute about 40 per cent of anthropogenic CO_2 emissions, and gas contributes about 15 per cent. OECD and other European countries account for about 70 per cent of global fossil fuel CO_2 emissions at present. However, energy consumption as a whole is the single largest contributor to greenhouse gas emissions in developing countries. Both India and China will be forced to increase coal-fired generation to meet the growing energy needs of their citizens for electricity through utilising domestic energy sources; some of this coal is low quality with enormous local, regional and, maybe, global impacts.

8.6   Investment costs in reducing dangers to health and environmental impacts

The modification of existing electricity-generation facilities to reduce emissions is achieved through the installation of emissions-control equipment that reduces SO_2, NO_x, and particulates. Emission controls add a significant amount to a plant's capital and operating costs. For example, in the United States pollution control costs are estimated at 40 per cent of capital and 35 per cent of operating costs [12]. For an entire pollution control system in a German power plant, SO_2 control represents 13 per cent of total the capital cost, NO_x control adds 6 per cent, and particulate control another 4 per cent. Control equipment in Japan constitutes an estimated 20–25 per cent of total capital costs for a new coal-fired plant, and 15–20 per cent of total electricity-generation costs. Add-on technologies of this nature do not change proportionately with the size of the facility and, as a result, pose relatively high costs to smaller facilities. Therefore their effects on cost of generating facilities in developing countries are larger. Generally speaking, pollution control for large electricity facilities will increase capital cost by 25 per cent and electricity production cost by up to 20 per cent; for smaller facilities the increase can be larger.

Particulate-control technology has a long history of use in power plants. The most cost-effective methods are electrostatic precipitators and fabric filters, both of which achieve 99.55 per cent particulate removal. Costs for both technologies are, for capital costs, about 4 per cent of total cost of plant and, for operating costs, about 5–8 per cent of total plant production cost. Fabric filters are especially relevant for low-sulphur coal use. Other methods include wet scrubbers, which have a high levelised annual cost, and mechanical collectors, which are not efficient enough (at 90 per cent removal) to meet most stringent environmental standards in developed countries.

In terms of electricity production, carbon-sequestration technologies are not commercial yet, as there are no developed and cost-effective pollution-control technologies that can reduce carbon emissions. However, natural-gas combined-cycle conversion eliminates sulphur emissions, reduces NO_x emissions by 90 per cent, and reduces carbon emissions by almost 60 per cent. In the United Kingdom, it has been estimated that it is cheaper to build and operate a combined-cycle gas-turbine

(CCGT) plant, as opposed to retrofitting an existing coal-fired plant. CCGT plants are more thermally efficient than conventional fossil-fired plants. They are also more energy efficient, and less costly, than the new 'clean coal' technology – fluidised bed combustion (which reduces sulphur and NO_x emissions without the installation of emissions control systems).

8.7 Evaluation of the environmental cost of electricity generation

Environmental impacts [13] of electricity generation were detailed in Table 8.1. The most serious impacts, which are difficult and very expensive to deal with, are air pollution damage in the form of emissions of SO_2, NO_x, particulates with air quality implications, and greenhouse gases, particularly CO_2. Such emissions will have a detrimental impact on health, buildings and similar properties, forests and climate change costs.

8.7.1 Assessing health impacts and costs

The major part of the dose of air pollutants inhaled by the general public is caused by road transport facilities and fuels and not by electricity production. There is remarkable evidence of linkages between very small particulate matter under $10\,\mu m$ diameter (PM_{10}) and mortality and detrimental health effects. Such very small particulate matter (PM_{10}) is normally produced by transport activities and not by electricity generation utilising coal, which produces the particulate matter of general dimensions.

There are many ways of valuing the impact [14] of air pollution on human health and well being. Willingness to pay (WTP) involves asking individuals questions about their willingness to pay, through a voluntary contribution or tax mechanism, to reduce air pollution to a safe level. A more direct method is that of *dose–response*, which involves finding a medical relationship between air pollution and observable health. Morbidity effects are valued by the cost of illness (COI) approach. This uses the cost of medical treatment and lost output as the social cost of the air pollution. However, the COI approach may be inferior to the WTP, since it ignores the disutility of illness.

In assessing the effect of air pollution on health, the detrimental health effects are caused by the small particulate matter (PM_{10}) emerging from car exhausts, rather than the usual general dimension particulate matter emitted from power-station chimney stacks. Utilising WTP methods, it is estimated that annual health costs of PM_{10} emissions in the UK, mostly from transport, may be something of the order of £14 billion [15]. This is far more damage than that can be caused by electricity production.

8.7.2 Assessing damage to buildings, properties and forests by air pollution

SO_2 and NO_x are the main air pollutants that are transboundary and can cause damage to buildings and similar properties, as well as to forests. In addition, the deposition of sulphur and nitrogen species causes oxidation of soils and fresh water. SO_2 emissions are mainly caused by firing fossil fuels, particularly for electricity generation. Brown coals, in particular, have a very high sulphur content and cause more emissions per unit of electricity generated than ordinary hard coal or other fossil fuels. SO_2 emission can be a serious cause of local and regional damage with significant financial cost.

Such emissions, and correspondingly damage cost, can be significantly reduced by either fuel switching or by add-on measures to abate emissions. Fuel switching involves using cleaner fuels. Natural gas (and LNG) is relatively benign with no SO_2 emissions and, if economically available, can significantly contribute to the reduction of air pollution. Utilisation of energy-efficient generation technologies and conservation can also significantly help. If fuels such as natural gas are not available, then the alternative for reducing air pollution is to utilise add-on measures. These are technologies and facilities that abate emissions, mostly of SO_2, NO_x and also particulate matter. Utilising these measures should be pushed to the extent that the cost of additional measures should be equal to their environmental benefits.

Sulphur reduction can be achieved prior to combustion by washing or by adding powdered limestone during firing, which can reduce the sulphur emissions by almost one third. However, major reductions are achievable after combustion by flue gas desulphurisation (FGD), by using limestone (which yields gypsum), or by regenerative techniques that yield sulphuric acid. Utilising all three techniques can reduce the sulphur emission by almost 95 per cent, but at enormous investment and operational cost as explained earlier. In order to assess the extent of these measures, it is essential to know two things: the marginal cost of abatement per each unit of sulphur and the cost of environmental damage that unit of sulphur causes. *Cost curves* for sulphur abatement can be plotted showing the extent of abatement against cost.

Assessment of environmental damage is much more diffucult [16]. Emissions are carried away over wide distances by complex weather patterns. Acidation takes place in dry and wet weather (acid rain), causing the damage described above. In the case of damage to building materials, the extent of damage depends on the extent of emissions and air pollution, and also on the type and quality of materials used. Stone buildings are less affected than brick buildings. The cost of the damage can be assessed by two methods: dose–response functions and the maintenance cycle approach.

In the dose–response functions method, the extent of damage is assessed by dose–response functions that are normally prepared for building stones and metals, where a critical damage level is defined for each material. The value of the national cost of the air pollution will equal the cost for replacement or repair of the damaged material.

The maintenance cycle approach starts with the same assessment of damage to materials by dose–response functions. However, it recognises that a reduction in emissions concentration will lead to a corresponding reduction in maintenance requirements of building material. Therefore, utilising the dose–response function, an estimation of the saving in maintenance is allocated for a certain reduction in emissions level through a corresponding investment. A value of total national savings is calculated for a given reduction in SO_2 concentration. Utilisation of either method has to be handled with care, taking into consideration the different factors other than SO_2 (like weather) that contribute to material damage.

To evaluate costs of damage of SO_2, a case of a 2 GW power station utilising 1.6 per cent sulphur coal, operating at an efficiency of 36 per cent, and not fitted with FGD facilities was considered. The cost adder in terms of p kWh^{-1} was computed. This indicated that cost of damage to buildings in the UK is put at 0.07p kWh^{-1}, within a range of 0.04–0.19p kWh^{-1}, depending on the assumptions utilised and the location of the power station with respect to urban areas [17].

A similar approach was used to assess damage by SO_2 to European forests. Damage functions were created to chart the annual loss in the value of wood (estimated by multiplying the estimate of volume damaged by a price of £23 m^{-3}) as affected by acid depositions in grams per square metre per annum. For the above described power station, estimated cost adders are 0.05 and 0.01p kWh^{-1} for SO_2 and NO_x, respectively [18].

8.7.3 Climate change costing

Climate change due to emission of greenhouse gases (CO_2, methane) is the most controversial aspect of air pollution. This is because of the existence of minimal evidence at present and the fact that damage can only be ascertained in the far future, 50 years or more. CO_2 emissions are the main source of greenhouse gases. These are gases that are supposed to accumulate in the atmosphere with a long residence time, leading to higher long-term atmospheric temperature.

Electricity generation was supposed to account, in 2002, for about one third of global greenhouse gases emissions from energy utilisation [2]. With increased growth of electricity generation, at a rate of almost 1.5 times that of total primary energy utilisation, electricity production will assume a large proportion of emission of gases, particularly CO_2 (37 per cent of global energy emissions in 2020), that may cause significant climate change in the future. Developing countries, particularly China and India, which have very high electricity demand growth potential, are dependent on local coal supplies that emit increasingly large amounts of CO_2 in the long term; although it is not easy to ascertain the extent and results of such emissions in the future, and correspondingly the extent of damage to human welfare and property. It is wiser to adopt the minimum regrets policy, which advocates the reduction in the rise of greenhouse gases emissions, and preferably their stabilisation as soon as possible, through energy efficiency, conservation and utilisation of relatively benign fuels like natural gas [19].

It is not possible to assess, with any degree of certainty, the future implications of increased emissions or concentration in the atmosphere of CO_2, and correspondingly to understand the mechanisms of global warming and its social and economical impact. It is, however, certain that whereas other air pollutants have local and regional impact, that of CO_2 is going to be global and can be much wider reaching. Whereas the impact of other air pollutants can be contained in a very significant way by technology (other than by fuel switching, efficiency and conversation), containment of CO_2 emissions is taking more time. Zero carbon emissions will be possible in the near future, but they may prove to be costly; correspondingly, only industrialised countries will be able to afford them (see also Chapter 9).

Climate change can have two sorts of impact. One is market related, which is reflected in monetary terms and national accounts. The other is non-market related in that it has an intangible effect, like impacts on human amenities and ecosystems. Such damage can usually be estimated by the WTP concept. However, such estimates are not always available for assessment of global warming impacts. Other indicators are utilised to assess the welfare impacts on climatic change such as reduction in revenues or returns from capital and land. The IPCC Working Group III (1995) [20] estimates that a future impact of an atmospheric CO_2 concentration of twice the pre-industrial level (a scenario called $2 \times CO_2$), will have the following damage costs:

- world impact: 1.5–2.0 per cent of world GNP,
- developed country impact: 1–1.5 per cent of national GNP,
- developing country impact: 2–9 per cent of national GNP.

Note that these figures include both adaptation costs and residual damages.

Such impacts are believed to vary from one country to another. Very few countries can have beneficial impacts. Asia and Africa, which accommodate most of the world's population, are likely to suffer extreme damage owing to severe life and morbidity impacts. However, these estimates are not comprehensive and are highly uncertain.

Greenhouse gases are stock pollutants; that is, an emission now will have long-term effects that stretch over several decades. What is important is to estimate the marginal cost of CO_2 emissions, that is, the discounted present worth extra damage for one extra ton of carbon emitted. To achieve this, there is a need to consider two scenarios: the first is the present value of future damages associated with certain emissions, and the second has marginally different emissions in the base period. The results of such an approach depend greatly on the choice of the discount rate and variations in scenario assumptions, with results varying between the wide margin of US\$5–US\$125 as a marginal cost of CO_2 emission (1990) per ton of carbon. Most estimates are on the lower margin. Future emissions will have a higher marginal cost owing to the stock effect [21].

The marginal social costs per ton of air pollutants emitted were valued by a commission of European Communities/United Stated (CEC/US – 1993), and the results are shown in Table 8.4. Such figures are useful in evaluating the economics of abatement facilities and measures, and utilising evaluation techniques described previously.

A study [22] of UK air pollution damage is demonstrated in Figure 8.2. The figure clearly depicts that SO_2, mainly from power generation, causes the greatest air pollution damage in the UK. However, over recent years, there has been a dramatic decrease in SO_2 emission. This resulted from increased public awareness and demand, which led to regulatory policies, switching from coal to cleaner fuels, efficiency measures, the wider refinement and adoption of emissions inhibition technologies.

It has been estimated [23] that the average damage costs per unit electricity generation from fossil power plants in EU-15, in 1990, amounted to 6.4 cents kWh^{-1}, this was more than the actual cost of production; since then there have been considerable improvements.

Table 8.4 Marginal social costs per ton of air pollutant emitted (US$) – 1993 [20]

	CO_2	SO_2	NO_x	PM_{10}
Health		1530	470	10 350
Forestry		1760	1220	
Materials/buildings		480	320	320
Climate change	7			
Total	7	3770	2010	10 670

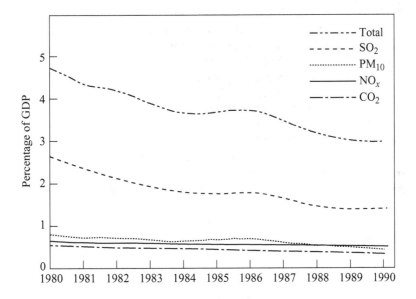

Figure 8.2 Cost of air pollution damage in the UK [22]

8.8 References

1 JARET, P., 'Electricity for Increasing Energy Efficiency', *EPRI J.*, April/May 1992, **17**, (3)

2 'World Energy Outlook', International Energy Agency (IEA) (OECD, Paris, 2000)

3 KHATIB, H.: 'Electrification for Developing Countries', *EPRI J.*, September 1993, **18**, (6)

4 'Foreign Market for U.S. Clean Coal Technologies', Department of Energy, DOE/FE-0317, Washington, DC, 1994

5 PESKIN, H. M., and LUTZE, E.: 'A survey of resource and environmental accounting in industrialized countries', The World Bank, Environment working paper 37, 1990

6 'Senior expert symposium on electricity and the environment', IAEA, Helsinki, Finland, May 1991

7 'Greenhouse gas emissions – the energy dimension', IEA (OECD, Paris 1991)

8 KREWITT, W.: 'External costs of energy', *Energy Policy*, August 2002, **30**, (10)

9 AWERBUCH, S. and DEEHAM, W.: 'Do consumers discount the future correctly?', *Energy Policy*, January 1995, **23**, (1)

10 'The economic appraisal of environmental projects and policies', OECD and Economic Development Institute of the World Bank, 1995.

11 PERRY, T.: 'Today's view of magnetic fields', *IEEE Spectrum*, 1994, **31**, (12)

12 MODRE, T.: 'Utility workers and EMF health risks', *EPRI J.*, 1995, **20**, (2)

13 KHATIB, H., and MUNASINGHE, M.: 'Electricity, the environment and sustainable world development', WEC Commission: Energy for Tomorrow's World plenary session, 8, World Energy Council 15th Congress, Madrid, 1992

14 CAITHROP, E., and MADDISON, D.: 'The dose response function approach to modeling the health effects of air pollution', *Energy Policy*, July 1996, **24**, (7)

15 PEARCE, D., and CROWARDS, T.: 'Particulate matter and human health in the UK', *Energy Policy*, July 1996, **24**, (7)

16 APSIMON, H., and WARREN, R.: 'Transboundary air pollution in Europe', *Energy Policy*, July 1996, **24**, (7)

17 APSIMON, H., and COWELL, D.: 'The benefits of reduced damage to buildings from abatement of sulphur dioxide emissions', *Energy Policy*, July 1996, **24**, (7)

18 GEOGORY, K., WEBSTER, C., and DURK, S.: 'Estimates of damage to forests in Europe due to emissions of acidifying', *Energy Policy*, July 1996, **24**, (7)

19 'Energy for Tomorrow's World', World Energy Council (Kogan Page Ltd., London, 1993)

20 'Externalizes of the fuel cycle: extern project', Commission of the European Communities/United States (CEC/US), working documents 1,2,5 and 9, Directorate General XII, European Commission, Brussels

21 FRANKHAUSER, S., and TOL, R.: 'Climate change cost', *Energy Policy*, July 1996, **24**, (7)

22 HAMILTON, K., and ATKINSON, G.: 'Air pollution and green accounts', *Energy Policy*, July 1996, **24**, (7)

23 KREWITT, W. *et al.*: 'Environmental damage costs from fossil electricity generation in Germany and Europe', *Energy Policy*, March 1999, **27**, (3)

Chapter 9

Electricity generation in a carbon-constrained world

9.1 Introduction

External factors, most importantly, the need to pursue the goals of sustainable development, have had a growing influence on the future of the ESI. International agreements like the Kyoto Protocol and the anxiety of governments to ensure energy security are modifying the way the business of electrical power is being conducted [1].

In May 2001, the US government formulated its energy policy. In February 2002, the UK government published 'The Energy Review'. This is a consultation paper, which recommends that the overall aim of future UK energy policy should be 'the pursuit of secure and competitively priced means of meeting our energy needs, subject to the achievement of an environmentally sustainable energy system'. In the short term it sees the expansion of the role of renewable generation and the promotion of energy efficiency as being the most cost-effective ways of meeting these policy aims and the Kyoto targets. In addition to the existing target of 10 per cent of electricity being generated from renewable resources by 2020, it also proposes a target of 20 per cent improvement in domestic energy efficiency by 2010, and a further 20 per cent improvement by 2020.

For the longer term, the review paper recognises the level of uncertainty involved in trying to predict too far into the future. Given that there is a strong chance that there will be even more stringent carbon-reduction targets post-2012, the review paper recommends retaining maximum flexibility by creating and keeping open all low-carbon energy options. These options include renewable electricity generation, small-scale combined heat and power (CHP), nuclear power and clean coal.

Most of the environmental achievements will have to be attained through more efficient production and utilisation of electrical power. A new factor has come into play: the ESI plans are increasingly becoming driven by governmental political considerations, as agreed in international forums.

9.2   Climate conventions and Kyoto

The most significant international environmental treaty ever drafted [2], the United Nations Framework Convention on Climate Change (UNFCCC), was agreed at the 1992 Earth Summit in Rio de Janeiro and sets out a framework for action to control and cut greenhouse gas emissions. A Protocol to the Convention was subsequently adopted in 1997 at the Third Conference of the Parties held in Kyoto, Japan. This provides a practical international process to reduce emissions that can contribute to global warming. Under the Kyoto Protocol, industrialised-country participants or parties agree to reduce their overall emissions of the six greenhouse gases (GHGs) by an average of 5.2 per cent below a 1990 baseline between 2008 and 2012, the so-called 'first commitment period' (see Section 9.2.1).

One of the primary targets for attaining carbon emissions reduction will be the power sector. Modern power stations, particularly those operating in a combined-cycle mode have very high efficiencies and use relatively benign fuels such as natural gas (see Figure 9.1). Such efficiencies, at nearly 60 per cent, are almost twice the efficiency of vintage available plants. Emissions from combined-cycle high-efficiency plant firing natural gas is only 40 per cent of that of a modern coal-firing thermal power plant (even less in case of brown coal). Shutting down an old coal-firing plant and replacing it by a modern CCGT plant will reduce carbon emissions by a factor of 3–3.5. Significant achievements have been made recently in higher CCGT plant efficiency (see Figure 9.2).

The power sector itself can expect to face various forms of fiscal tightening to curb emissions, such as taxes levied on energy used as well as additional and tougher regulatory targets for energy efficiency. The power sector will also be encouraged to

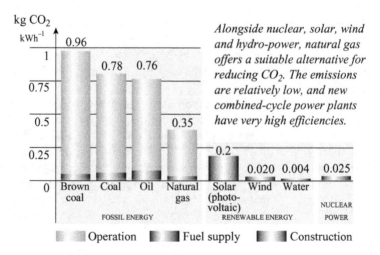

Figure 9.1   Comparison of CO_2 emissions [3]

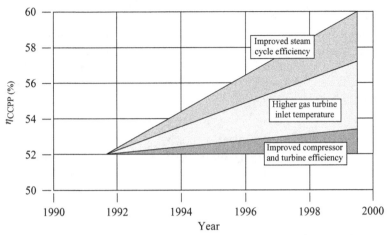

During the past decade, the efficiency, i.e. the energy utilisation, of combined-cycle power plants has been improved by several percentage points. Improvements in steam turbines and an increase in gas turbine inlet temperature were the key factors behind this gain.

Figure 9.2   Combined-cycle power plant efficiency development [3]

participate in the so-called Kyoto 'flexible mechanisms', a raft of activities permitted under the Protocol to curb emissions.

9.2.1   Case Study: the Kyoto Protocol and the Marrakech declaration and accords

Objective

The so-called Annex B countries – all signatory countries that have set absolute emissions goals for the six most important GHGs – have committed themselves to reducing their average emissions by 5.2 per cent relative to 1990 levels by 2012. Germany aims to reduce greenhouse gas emissions by 21–25 per cent, which equals 212 million tons of CO_2 equivalents.

Ratification

The Kyoto Protocol was planned to go into effect immediately after the World Summit on Sustainable Development in Johannesburg in September 2002, if a minimum of 55 countries that accounted for at least 55 per cent of CO_2 emissions in 1990 had ratified the agreement by that time.

Approved instruments for attaining emissions objectives:

1   Direct national reduction of GHGs
2   Clean Development Mechanism (CDM)
3   Joint Implementation (JI)
4   International Emissions Trading (IET).

Verification and sanctions

All countries have committed themselves to regular publication of figures on greenhouse gas emissions and reductions achieved. A commission will supervise compliance. If commitments are not fulfilled, sanctions may be enacted. For every ton of the CO_2 emissions target not achieved by 2012, an additional reduction of 1.3 tons must be achieved later.

Assistance for developing countries

Several funds for implementing environmentally friendly policies in developing countries have been set up. Developing countries are not required to implement concrete climate protection measures until 2012.

9.3   Electricity generation and climatic change agreements

Electricity generation is going to be the primary target for emissions reductions as agreed to by climatic change agreements. This is because of the following points.

- An increasing share of final energy is being utilised in the form of electricity.
- Prospects for curbing emissions, through improved efficiency, cleaner fuels, renewable and nuclear power and similar means, are easier and with more significant results in the case of power generation than in other energy conversion processes. New power-generation technologies will significantly assist in this regard.
- Emissions from power generation are concentrated in one place – the power station – where it is easier to deal with.
- Electricity lends itself more easily to project mechanisms that allow for emission reductions through 'emissions trading' and similar flexible mechanisms.

The growing importance of global CO_2 emissions from power generation are demonstrated in the following Table 9.1, adopted from IEA projection.

Table 9.1   Electricity contribution to CO_2 emissions

	1971	1977	2010	2020
Global CO_2 emissions (Mt) from fossil fuels	14 753	22 984	30 083	36 680
Emissions from power generation (Mt)	3885	7663	10 671	13 479
Electricity/global (%)	26.3	33.3	35.8	36.8

Source: IEA – WEO 2000

Table 9.2   Carbon emissions forecasts for the UK

	Carbon emissions in 2010 (million tons)	Percentage change between 1990 and 2010
Power generation	45.1	−16.1
Energy-intensive industries	11.6	−21.0
Other industries	12.6	−15.1
Road transport	30.4	1.7
Air transport	1.9	22.7
Households	25.6	18.8
Commerce, etc.	9.9	5.1
Total	**149.5**	**−6.4**

From Table 9.1 it is clear that power generation is assuming an ever-greater share of global CO_2 emissions in spite of improvements in generation efficiency and an increasing share of natural gas in the process. Correspondingly any attempt to curb global emissions will need to have power generation at its centre of interest.

In the UK it is expected that carbon emissions from power generation, in 2010, will account to 45 million tons, i.e. 30 per cent of UK emissions in that year. This is 16.1 per cent less than 1990 figures, and almost three times the national average, signifying the potential for the power industry to reduce carbon emissions as detailed in Table 9.2.

9.3.1   Flexible project mechanisms

These are designed to enable countries to deliver part of their committed emissions targets – 'Assigned Amounts' – through investment in projects that reduce emissions in another country. Power generation is likely to be the most flexible vehicle for such projects implementation. There are two project-based mechanisms.

- JI allows for investment by one developed country in a project that reduces GHG emissions in another developed country (like a Western European country investing in a country in Eastern Europe). 'Emissions Reduction Units' (ERUs) generated by such projects can be used to meet the investing country's commitments.
- The CDM is designed to bring development to developing countries through projects that result in a reduction in GHG emissions. 'Certified Emission Reduction' (CERs) units are generated that can be purchased by a developed country and used to meet its commitments (Figure 9.3).

To qualify under the CDM, projects put forward for registration to the UNFCCC Secretariat's executive board of the CDM must meet a range of sustainability and environmental criteria set down under the Kyoto Protocol. One of these is the requirement for a verified project emissions baseline.

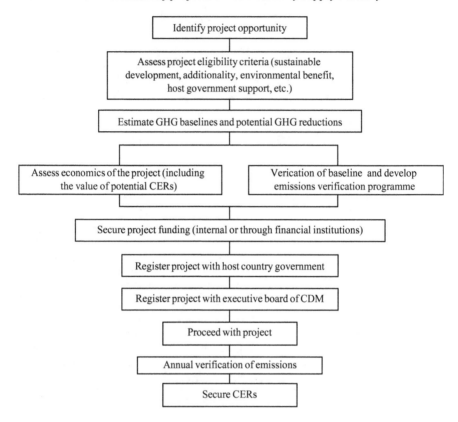

Figure 9.3   The CDM process [2]

Likewise, under a verification programme specified in the registration process, independent operational entities (third-party verifiers) will periodically (probably annually) verify GHG emissions from the project to enable them to recommend to the executive board of the CDM the number of emissions credits (CERs) that can be issued.

The Kyoto mechanisms offer the power-generation sector an incentive to invest in emissions-reduction projects since ERUs and CERs should be funded with permits issued as 'Assigned Amounts' that can subsequently be monetised through emissions trading. The potential size of the CDM market remains uncertain (especially given the uncertainty over supplementary). But the best forecasts made to date suggest it will be between 700 and 2100 million tons of CO_2 per year.

9.3.2   Emissions trading

This market mechanism is widely seen as the best way to help businesses identify the most cost-effective options available to reduce emissions. The trading market should stimulate the development of new technology, because it sets a real cost on emissions and provides market incentives to reduce that cost.

Several countries, for example, the UK and the Netherlands, are developing national trading systems for GHGs. Trading in atmospheric emissions is not new; both sulphur dioxide (SO_2) and nitrogen oxide (NO_x) emissions have been successfully traded in the US for many years. Emerging systems are based on a 'cap and trade' concept where participants stay within a basic allocated emissions allowance. Where actual emissions of the business are above its total allowance for a given year, then it must buy more allowances from the market to ensure compliance. Where emissions are below an allocation, a company can sell allowances. The mechanism for setting allowances requires a 'base year' for which verifiable data can be reported and against which annual allocations can be set. Successful GHG emissions have to be accurate, reliable and independently verified by a qualified third party.

Case study: UK Emissions Trading Scheme

The UK market officially launched this scheme in April 2002, but forward allowance trades were carried out from September 2001 at prices between £4 and £6 per tonne of CO_2. Since the scheme was launched, many more trades have taken place at widely varying prices, up to £6 per tonne.

Participation in the UK Emissions Trading Scheme [4] is open to most UK companies and is voluntary. Electricity generators, the transport sector, households and large landfills are not currently eligible to participate but may be able to enter the scheme through emissions-reduction projects once the rules for this are established. The government has provided significant incentives for companies to participate in the form of a rebate on the Climate Change Levy (CCL) for energy-intensive companies and payments of £53.37 per tonne of CO_2 reduced for companies that entered through an auction mechanism in March 2001. More than 6000 companies are eligible to trade.

The UK scheme is a pilot, running from 2002 to 2006, and so it will still be in operation as the EU scheme begins. The UK scheme is complicated by its interaction with the CCL. Many companies that receive the CCL rebate have rate-based targets and a 'gateway' has been established between the absolute sector and the rate-based sector to avoid inflation of the absolute sector from the rate-based sector. The gateway will close whenever aggregate sales from the rate-based sector to the absolute sector equal the sales in the other direction.

An electronic registry records the holdings of allowances for each participant, and tracks allocations, transfers, and final cancellation or retirement of allowances. Initially the registry will serve only the UK scheme, but in time it is envisaged that it will act as the UK's national registry for international trading under the Kyoto Protocol. By 2008, the registry will need to interface will overseas registries and will become the main tool for tracking changes to UK-allowance holdings. Ultimately, the government envisages that participants will be able to use any allowances or credits obtained through the Kyoto mechanisms subject to the appropriate approval.

9.3.3   Encouraging emissions reductions

In the UK, to raise the cost of energy use, the CCL is imposed on business and public sector energy use from April 2001 whereby only good quality CHP plants and renewable energy (other than hydro) of more than 10 MW capacities will be exempt. Revenues from the CCL will be recycled to business via a 0.3 per cent cut in employers taxation and £150 million ($219 million) of support for energy-savings measures.

Energy-intensive sectors may also win an 80 per cent discount in the CCL under special arrangements in the form of climate change agreements (CCAs) with the government. These agreements set tough targets for participants to improve energy efficiency and/or reduce emissions. They do not dictate investment in specific technologies or techniques and they permit sectors to achieve their targets either by trading emissions allowances with other companies within their CCA or by participating in a wider national emissions-trading scheme.

9.4   Introduction to low emissions and emissions-free power generation

Carbon sequestration is one of three critical elements in a three-pronged approach to alleviate global climate change concerns. The other elements involve increasing the efficiency of energy generation and use, and reducing the carbon content of fuels (e.g. natural gas and renewable and nuclear energy).

Carbon sequestration is an important option because it does not require a massive overhaul of the world's energy infrastructure. The goal is to develop sequestration approaches that cost $10 or less per ton of carbon removed equivalent to adding 0.2 cents kWh^{-1}. Empirical evidence from advances in sodium carbonate technology suggests that this goal is realistic, indicating 50 per cent CO_2 capture from a power plant at $15 per ton of carbon removed. In the US as well as Europe carbon sequestration programmes are underway on several fronts including carbon capture and separation, sequestration in geologic formations/oceans/terrestrial ecosystems and advanced concepts.

Nuclear-power generation and generation from most renewable sources are free from carbon emissions. This should encourage their utilisation, particularly with the prospect of carbon taxation, but with the following provisos.

- Nuclear energy is increasingly becoming unpopular, particularly in Western Europe, owing to antagonistic public opinion, whether this is justified or not. Also the high front investment cost, long lead times and prospects of future legislation renders nuclear energy unfeasible for private utility investors.
- Renewables are generally weather dependent and as such their likely output can be predicated but not controlled. The only control possible is to reduce the output below that available from the resource at any given time. Therefore, to safeguard system stability and security, renewables must be used in conjunction with other,

controllable, generation and with large-scale energy storage. There is a substantial cost associated with this provision.

- Countries opt for energy security. Such security is enhanced by utilisation of local resources, particularly local cheap coal, as is the case in North America, South and East China. These countries will continue to rely on their local coal resources, increasingly with clean-coal technologies, rather than resort to other options.

Future prospects of fossil fuel, nuclear as well as renewable energy for carbon-free generation, including cost as well as technologies, are briefly summarised in Section 9.4.1.

9.4.1 Generation cost of power sources

In the EU, research was undertaken to assess the cost of the socio-environmental damage associated with every type of electricity generation [5]. Such costs are not normally charged to the power producer and are not passed to the consumer. But such 'hidden' costs are increasingly being taken into consideration by regulatory authorities in their pursuit of sustainable strategies. These are the costs estimated to be borne by the society in negative health impacts, resource depletion, global warming, etc.; an estimation of the 'clear' production costs, as well as the 'hidden' social costs are given in Table 9.3.

A relatively wide range of costs is indicated for each technology due to peculiarities of individual projects, risks and discount factors and individual financial costs involved.

9.5 Clean fossil-fuel-based power technology

Programmes are underway to develop power plants of the future [6,7], using coal and natural gas as the primary fuel, that:

- are essentially pollution free;
- nearly double the current generating efficiency;
- have the capability to produce a varied slate of co-products, such as chemicals, process heat, and clean fuels.

Gasification technologies are an essential part of this effort because they provide the means to vastly expand the clean fuel base to more-abundant and lower cost resources (including biomass and wastes) and enable co-production of power and high value by products.

Gasification breaks feedstocks down into basic constituents so that pollutants can be readily removed and clean fuels or chemicals can be produced.

- Resultant clean fuels can be used to power highly efficient and clean gas turbines, fuel cells, and fuel cell/turbine hybrids.
- Gasification concentrates the CO_2 constituent, which facilitates capture and recycle or sequestration.

Table 9.3   Total cost of electricity generation

Item	Power source	Costs					
	Description	Production $p\,kWh^{-1}$		Social $p\,kWh^{-1}$		Total $p\,kWh^{-1}$	
1	Conventional coal fired (with FGD and CO_2 abatement)	4 — 5		2.5 — 4.5		6.5 — 9.5	
2	Oil-fired (with CO_2 abatement)	5 — 6		2 — 3		7 — 9	
3	Combined cycle gas turbine	1.2 — 3		0.6 — 1.2		1.8 — 4.2	
4	Micro-combined heat and power	— — —		— — —		— — —	
5	Nuclear, AGR and PWR	3 — 3		0.15 — 0.15		3.15 — 3.15	
6	Hydro	1.5 — 2.5		0.4 — 0.6		1.9 — 3.1	
7	Tidal power	2 — 6		0.4 — 0.6		2.4 — 3.6	
8	Onshore wind	2 — 3		0.1 — 0.1		2.1 — 3.1	
9	Offshore wind	5 — 5		0.1 — 0.1		5.1 — 5.1	
10	Landfill gases	2 — 3		0.5 — 0.5		2.5 — 3.5	
11	Municipal incineration	1 — 2		0.6 — 0.6		1.6 — 2.6	
12	Biomass, field, forest, straw	1.5 — 4		0.6 — 0.6		2.1 — 4.6	
13	Import	1.7 — 1.7		0.15 — 0.15		1.85 — 1.85	
14	Diesel	4 — 6		0.5 — 0.5		4.5 — 6.5	
15	Photovoltaic	8 — 10		0.2 — 0.6		8.2 — 10.6	
16	Hydro-pumped storage	— — —		— — —		— — —	
17	Electrochemical storage	— — —		— — —		— — —	

Source: Reference [5].

- A clean-coal technology programme has sponsored integrated-gasification combined-cycle (IGCC) projects, representing a diversity of gasifiers and clean-up systems that are helping to pioneer the commercial introduction of this next generation power concept.

Gas separation membranes are important for enhancing the cost and performance of technologies on many fronts, by offering the promise of displacing energy-intensive, costly cryogenic and chemical means of separating out selected constituents from a gas stream, such as oxygen, hydrogen, CO_2, or pollutants.

- Separation membranes have the potential significantly to enhance the cost and performance of gasification-based technologies.
- Ion transport membranes for oxygen separation are in an advanced stage of development.
- In addition to gasification applications, ion transport membranes open the possibility of combustion with oxygen rather than air to eliminate nitrogen-based pollutants and to concentrate the CO_2 constituent, which enables capture.

- High-temperature ceramic membranes hold the promise of making a hydrogen-based economy feasible and have immediate application to fuel cells, and fuels and chemical production.
- CO_2 separation membranes have the potential to significantly reduce the cost of CO_2 capture, which is the most costly facet of the carbon sequestration process.

Advanced gas-turbine development is essential because gas-turbines will be a mainstay in the power-generation industry for the foreseeable future, operating on natural gas in the near- to mid-term and on gasification-derived synthesis gas in the longer term. Advanced turbine systems programmes for the future aim:

(i) to improve the efficiency and overall performance of the smaller gas-turbines in electric-generation service that are subjected to highly cyclic loads, and

(ii) to link these gas-turbines to fuel cells to push efficiency and environmental performance even higher.

Fuel cells have the potential to revolutionise power generation because they:

(i) operate on a range of hydrogen-rich fuels (natural gas, methanol, and gasification-derived synthesis gas),

(ii) represent a bridge to a hydrogen economy,

(iii) offer inherently high efficiency and are essentially pollution free (emitting water, CO_2 and heat),

(iv) lend themselves to distributed generation applications because of low emissions and quiet operation (owing to few moving parts).

- Distributed generation alleviates electricity transmission and distribution constraints (enhancing the reliability of electrical grids), enables combined heat and power applications (boosting thermal efficiencies upwards of 80–90 per cent), and offers an option to central-power generation.
- Phosphoric acid fuel cells are offered commercially and are penetrating niche markets, with efficiencies of 40 per cent – a few 250 kW units have already been sold.
- Second-generation, high-temperature molten carbonate fuel cell and solid oxide fuel cells are poised to enter the market, with efficiencies approaching 60 per cent when operated in a combined-cycle mode that uses the process heat to produce steam.
- Fuel cell/turbine hybrids that has the potential to raise efficiencies up to 70 per cent by using the process heat from high-temperature fuel cells to drive a gas turbine are under development. Fuel cell energy announced the start-up of a 250 kW molten carbonate fuel cell/capstone micro-turbine system, and a solid oxide fuel-cell-based system is soon to follow.

Fuel cells are unlikely to take the power markets by storm. As long as a developed hydrogen infrastructure is not available (and it may take decades), fuel cells are just another fairly efficient means of converting gas into power and heat; no more, no less. To succeed, manufacturing costs must first drop to a level that allows prices barely to exceed the $1000 \, \text{kW}^{-1}$ threshold. This may take quite some time.

9.6   Prospects for non-fossil-fuel power generation

9.6.1   Prospects for nuclear energy

Nuclear energy [8] still accounts for almost 16 per cent of global electricity production. However, for the last quarter of a century no nuclear plants were built in the US and very few were built elsewhere. Public awareness and opposition to new nuclear plants is no doubt the most significant factor in the recent demise of nuclear power. However, alarm about carbon emissions and the emissions-free character of nuclear power rekindled hopes for a revival in such facilities. In the first half of 2002, Finland announced a decision to proceed with its fifth nuclear reactor, the first such decision in Europe for a long time.

Nuclear power has many positive attributes that make it appealing to a balanced electricity-generation mix. These include a very large potential energy resource, demonstrated performance in maintaining very high levels of safe operation and public and workforce health and safety, no atmospheric pollution, low and stable fuel cost, and very high reliability. The substantial capital cost for new nuclear plant capacity, however, remains a significant, negative attribute of nuclear power.

This penalty cost of nuclear power has been aggravated by the rising efficiency and continuous fall in the cost of a modern CCGT plant. Recent cost assessment of an Advanced Light Water Reactor (ALWR) plant put the average cost of construction at about $1350 \, kW^{-1}$. For new nuclear capacity to compete economically with natural-gas-fired plants, the nuclear plant would need to have achieved a construction cost of possibly as low as $1000 \, kW^{-1}$.

Significantly higher natural gas prices, and/or the imposition of a hefty carbon tax will improve nuclear power plant's competitiveness. However, for nuclear, the potentially large consequences of a beyond-design accident, and the long-term impacts of radioactive waste are the main drivers that led to decisions for a moratorium or the phasing out of nuclear power in some European countries. The problem of proliferation also remains a critical issue.

9.6.2   Prospects for wind power

Wind power [9] is the most rapidly growing renewable energy source. Demand for it is expected to increase by 13 per cent a year over the next 20 years.

The cost of wind power is high compared with fossil fuels, but declining capital costs and improved performance are likely to reduce generating costs. Wind is expected to be competitive with fossil-fuel-based generation on the best sites on land over the next decade. Figure 9.4 shows how electricity-generating costs for wind power and fossil fuels may evolve in OECD Europe over the next 20 years.

Large land requirements and competition among different land uses could constrain growth. The intermittence of wind power and the unsightliness of wind turbines could further limit site availability. The effects of intermittence must be taken into consideration at the early stages of wind-power development, as the effects may become more obvious with higher shares of wind in the electricity mix.

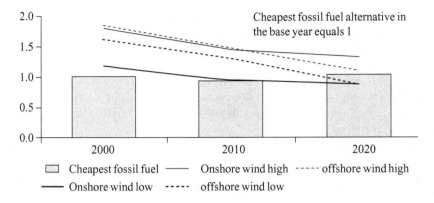

Figure 9.4   OECD Europe electricity generating costs for wind and fossil fuels [9]

9.6.3   Prospects for solar cells

In 2001, the world solar-cell production [10] soared to 395 MW, up 37 per cent over 2000. This annual growth in output, now comparable in size to a new power plant, is set to take off in the years ahead as production costs fall. Cumulative solar cell or photovoltaic (PV) capacity exceeded 1840 MW at the end of 2001.

The cost of electricity from solar cells remains higher than from wind or coal-fired power plants for grid-connected customers, but it is falling fast owing to the economies of scale as rising demand drives industry expansion. Solar cells currently cost around $3.50 W^{-1} for crystalline cells, and $2 W^{-1} for thin-film wafers, which are less efficient but can be integrated into building materials. Industry analysts note that, between 1976 and 2000, each doubling of cumulative production resulted in a price drop of 20 per cent. Some suggest that prices may fall even more dramatically in the future.

The European Photovoltaic Industry Association suggests that grid-connected rooftop solar systems could account for 16 per cent of electricity consumption in the 30 members of the OECD by 2010. If the costs of rooftop PV systems fall to $3 W^{-1} by the middle of the present decade, as projections suggest, the market for residential rooftop solar systems will expand. In areas where home mortgages finance PV systems and where net metering laws exist, demand could reach 40 GW, or 100 times global production in 2001.

More than a million homes worldwide, mainly in villages in developing countries, now get their electricity from solar cells. For the 1.7–2 billion people not connected to an electrical grid, solar cells are typically the cheapest source of electricity. In remote areas, delivering small amounts of electricity through a large grid is cost-prohibitive, so people who are not close to an electric grid are likely to obtain electricity from solar cells. If micro-credit financing is arranged, the monthly payment for PV systems is often comparable to the amount that a family would spend on candles or kerosene for lamps. After the loan is paid off, typically in two to four years, the family obtains free electricity for the remainder of the system's life.

PV systems provide high-quality electric lighting, which can improve educational opportunities, provide access to information, and help families be more productive after sunset. A shift to solar energy also brings health benefits. Solar electricity allows for the refrigeration of vaccines and other essentials, playing a part in improving public health. For many rural residents in remote areas, a shift to solar electricity improves indoor air quality. Photovoltaic systems benefit outdoor air quality as well. The replacement of a kerosene lamp with a 40 W solar module eliminates up to 106 kg of carbon emissions a year.

In addition to promising applications in the developing world, solar energy also benefits industrial nations. Even in the UK, a cloudy country, putting modern PV technology on all suitable roofs would generate more electricity than the nation consumes in a year. Recent research surrounding zero-energy homes, where solar panels are integrated into the design and construction of extremely energy-efficient new houses, presents a promising opportunity for increased use of solar cells.

Continued strong growth suggests that the solar-cell market will play a prominent role in providing renewable, non-polluting sources of energy in both developing and industrial countries. A number of policy measures can help to ensure the future growth of solar power. Removing distorting subsidies of fossil fuels would allow solar cells to compete in a more equitable marketplace. Expanding net metering laws to other countries and the parts of the US that currently do not have them will make owning solar home systems more economical, by requiring utilities to purchase excess electricity from residential solar systems. Finally, revolving loan funds and other providers of micro-credit are essential to the rapid spread of solar-cell technologies in developing nations.

9.6.4   The virtual power plant

Different generating facilities are likely to proliferate in the future. Beside conventional plants, these include large wind farms, biomass power plants, district heating, fuel cells, and PV plants. Connecting money energy suppliers together to form a virtual power plant [11] poses a special challenge for information and telecommunications technology. Numerous setpoint and actual values must be compared in a decentralised energy-management system, automation units have to be controlled, and forecasts of sun, wind and consumer behaviour must be obtained. Prospects for a virtual power plant will also be discussed in later chapters.

9.7   References

1 JACKSON, K.: 'Moving towards an energy policy for the UK', *Power Engineering Journal*, April 2002, **16**, pp. 50–51
2 EDDIES, M.: 'Climate Convention and Kyoto' *Power Economics*, February 2001, pp. 23–25
3 ASCHENBRENNER, N.: 'Gentle Revolution', *Siemens Magazine for Research and Development*, Spring 2002, pp. 50–54

4 ECOAL, World Coal Institute, **42**, June 2002, p. 5

5 NEWTON, M., and HOIEWELL, P. D.: 'Cost of sustainable electricity generation', *Power Engineering J.*, April 2002, **16**, (2) pp. 68–74

6 SMITH, C. M.: 'Technology and Fuels for Tomorrow', a lecture delivered in the meeting of the CFFS Committee 18th World Energy Congress, Argentina, October 2001

7 MCKEE, B.: 'Solutions for the 21st Century – Zero Emissions Technologies for Fossil Fuels – Technology Status Report', IEA, Paris, May 2002

8 EPRI Journal Online: 'Nuclear Power in America's Future', April 2002 (www.epri.com/journal)

9 ARGIRI, M., and BIROL, F.: 'Renewable Energy', *Oxford Energy Forum*, **49**, May 2002, pp. 3–5

10 FISCHTOWITZ-ROBERTS, B.: 'Sales of Solar Cells Take Off', Eco-Economy Update 2002–8, Earth Policy Institute 2002

11 MULLER, B.: 'The Power of Small Step', *Siemens Magazine for Research and Development*, Spring 2002, pp. 61–63

Chapter 10

Economics of reliability of the power supply

10.1 Introduction

Reliability of the power system is generally defined as the overall ability of the system to perform its functon [1]. It can also be defined as the ability of the power system to meet its load requirements at any time. Two distinct aspects of system reliability are identifiable: system security and system adequacy. System security involves the ability of the system to respond to disturbances arising internally, whereas system adequacy relates to the existence of sufficient facilities within the system to satisfy the customer-load demand [2].

As mentioned in Chapter 1, reliability of the electricity supply is of paramount importance. Interruptions (even transient supply problems) can cause serious financial and economical (social) losses. Electricity shortages can affect any country in two ways: they can handicap productive activities and seriously affect consumers' welfare. From the productivity standpoint, electricity shortages discourage investors by affecting production, increasing its cost, and requiring more investment for on-site electricity production or standby supplies. For small investors, the cost of operation is increased, since electricity from small private generation is more expensive than public national supplies. Electricity interruptions at home also cause consumers great inconvenience, irritation, loss of time and welfare [3].

To any economy, an unreliable power supply results in both short- and long-term costs. Costs are measured in terms of the loss of welfare and the adjustments that the consumers (such as firms), facing unreliable power supplies, undertake to mitigate their losses.

Service interruptions may trigger loss of production, costs related to product spoilage, and damage to equipment. The extent of these direct economic costs also depends on a host of factors such as advance notification, duration of the interruption, and timing. The latter refers to the time of day or season and to the prevailing market conditions regarding the demand for the firm's output. These direct costs can be very high, particularly under conditions of uncontrolled load shedding and transmission and distribution outrages, i.e. sudden interruptions in service without advance notification. In addition to the direct costs noted earlier, there are indirect costs to the economy because of the secondary effects that arise as a result of the interdependence between one firm's output and another firm's input.

Chronic electricity shortages and poor reliability of supply trigger long-term adjustments. If a firm's expectations are that shortages and unreliable service will persist, then they will respond in one or more ways. The installation of back-up diesel generator sets is the most common long-term adjustment taken by commercial consumers and small industrial firms. It has been estimated that a substantial amount of the total installed generating capacity in many developing countries (in the order of 20+ per cent) is in the form of standby generation on customer premises [4].

Shortages of electric power and supply interruptions occur because of either of the following two reasons.

(i) Unreliability of the supply due to the non-availability of generating plants, or breakdowns in the transmission and distribution system. Such unavailability can occur in varying degrees in any power system in the world: *system security*.

(ii) Shortfalls of delivered electric power even under the best conditions of the electric system. Such shortfalls usually occur in developing countries. Most of these suffer capital shortages owing to inadequate number of generating facilities capable of matching the peak demand, and with limitations in the transmission and distribution system, particularly to rural areas: *system adequacy*.

The ability of a power system to meet demand and deliver adequate electric energy to the consumers, is termed above as system adequacy. To provide such adequacy and to overcome supply shortages and interruptions, there is a need for investment. Most investments in the ESI are meant to reduce the prospect of shortages, and maintain

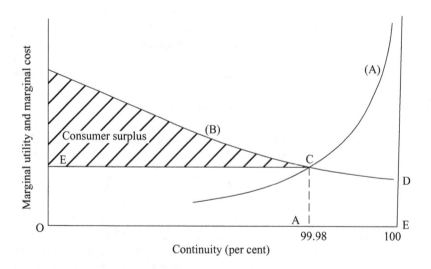

Figure 10.1   Marginal utility and marginal cost of electric availability

and improve reliability. Most of the shortages can occur as a result of the growth in the demand, which necessitates generation extension and network strengthening. However, and in spite of large investments, interruptions are inevitable. Costs of improving continuity of the supply can be very high after a certain level of reliability. As shown in Figure 10.1, such costs can increase at an exponential rate beyond a certain level of reliability [5].

The function of the power system is to provide electrical energy as economically as possible and with an acceptable degree of reliability and quality. Economics of reliability tries to strike a reasonable balance between cost and quality of service. Such balance varies from one country to another, just as from one category of consumers to another. Most developing countries cannot afford the high costs of ensuring the high reliability of supply, taken for granted in industrialised countries. Gradually, however, quality of supply is improving in developing countries.

10.2 The value of reliability

By value of reliability, we mean the estimation (in monetary terms if possible) of the benefits and the utility derived by the consumer from achieving (or improving) a certain level of availability of the supply. The cost of unreliability over a certain period is the social cost (financial and economic) suffered by consumers as a consequence of supply interruption during that period. Cost–benefit analysis of reliability economics in planning of the power system has not advanced in the way it has in some other public services, e.g. transport. This is because of the great difficulty of obtaining comprehensive assessment of the reliability of supply from the generating system down to the consumer terminals. Even more difficult is the evaluation of consumer financial and economical loss, and the inconvenience due to interruptions [6].

The electricity tariff does not reflect the benefits that consumers (the economy) derive from the existence of the electricity supply. In Figure 10.1, the tariff, if it is fixed at a long-run marginal cost, is equal to CA. What the consumer pays for is the annual quantity of consumption multiplied by the tariff (area OECA). The benefits are the entire area under curve (B) and to the left of CA. This is much higher than what is paid. The difference is called the *consumer surplus*, which is the extra benefit that the consumer derives out of the presence of the supply over what is paid in tariffs, which is usually substantial. The consumer loss of utility due to interruption is depicted by area ACDE.

It is possible, however, to assess the contribution of different schemes towards the continuity of supply and to choose the one with the maximum contribution within economical cost. The problem still remains of what is the 'economical cost'? Because of these difficulties, empirical design rules and national standards for supply security have been set to guide and limit the planners, without resorting into actual monetary valuation of reliability benefits. In a few countries, however, monetary values have been fixed for evaluating the cost of loss of supply. These values are utilised to evaluate cost of interruptions and compare these with the system strengthening cost, in order to justify its execution.

What the planner would aim at is not a completely uninterruptible supply, the cost of such a system would be prohibitive as shown by curve (A) of Figure 10.1. The aim should, however, be the point of 'optimum reliability' point C in the figure. Point C is the intersection of the reliability marginal cost curve to the electricity supply industry (curve (A)) with the marginal cost of interruptions to the consumer (curve (B)). The intersection point is where the optimum reliability is obtained. It is where the marginal utility of the extra increment of reliability improvement to the consumer is equal to the marginal cost spent in achieving it by the supply industry.

It is possible to draw curve (A) over a wide range. This curve, which can be evaluated by modelling, is the least-cost system strengthening scheme as chosen from many schemes, which would lead to the same reliability but involve higher cost. It is, however, very difficult to draw curve (B). It is now possible, through studies and consumer surveys to obtain information and build tables of interruption cost data applicable to certain categories of consumers and certain areas [7]. Such data are valuable in estimating the value of reliability to assist in cost–benefit analysis for system strengthening.

10.3   Reliability in system planning

In order to understand the value of this approach, it is advantageous to review the existing methods that guide reliability planning and are applied by most of the electricity supply utilities worldwide. These can be summarised as follows [5]:

- empirical planning rules,
- supply design standards,
- simplified cost–benefits analysis, and
- detailed financial and economic evaluation.

The first three methods will be discussed in Sections 10.3.1–10.3.3, and the fourth method will be discussed in detail in Section 10.4.

10.3.1   Empirical planning rules

In this case, the planner, based on his experience and practice in dealing with similar situations, decides on the desired reliability level. This is achieved through the evaluation of the importance of the system and network and correspondingly the extent of redundancy to be incorporated, taking into account the following: the part of the system it is dealing with, voltage level and nature of the network, number and category of consumers, financial constraints, and past reliability records and past experience in handling a similar situation.

Almost all systems that do not have drawn out and written supply design standards, or probabilistic planning, employ such criteria. The empirical rules for generation involve a percentage reserve-margin method, without a loss of load expectation (LOLE) calculation; or for a small system, a firm generating capacity with the biggest set(s) out-of-commission. For the main network the '$n - 1$' rule is employed, which means that the loss of a main line from *n* parallel lines, or a transformer of *n* transformers, should not affect continuity of supply, with more emphasis being placed on important networks. In the distribution, and lower-voltage networks, 'rules of thumb' are employed in accordance with the consumer category and investment constraints mentioned earlier.

10.3.2 Supply design standards

Supply design standards are a step forward from the empirical approach, where the amount of past experience, performance, economical limitations and individual separate rules are reduced to a set of detailed guidelines for utilisation by planners to reliably design the system.

In generation, this involves a LOLE target, which stipulates that in any year the probability of capacity shortage should not exceed a certain value; usually a fraction of a day. Correspondingly, plans are drawn out in advance for the generation system strengthening so that LOLE would not exceed a certain standard. In the network, consumer groups and supply areas are classed in accordance with their demand, the minimum number of circuits available, and the target time for restoration of supply is specified. For interruption duration not to exceed a certain level, the network may need strengthening.

10.3.3 Cost–benefit analysis

Generally, not all cost–benefit analysis of the power system should necessarily involve monetary valuation of the cost of interruption. An increasing amount of work is being undertaken to evaluate the cost of different schemes and the probable amount of interruption. Further engineering judgement is used to choose the right plan, within the constraints mentioned, utilising empirical rules.

A simple actual example commonly encountered is that of choosing the method of protection of rural single feeders (Figure 10.2). Three methods are discussed and costed: expulsion fuses (EF), autoreclose circuit breakers (AR) and AR with automatic-sectionaliser (AS), in the middle of the line. The cost and predicted continuity performance of these schemes when applied to a particular rural network are summarised in Table 10.1.

Cost–benefit engineering judgement can now be applied. The employment of EF with an expenditure of only £500 on network protection will involve 12 000

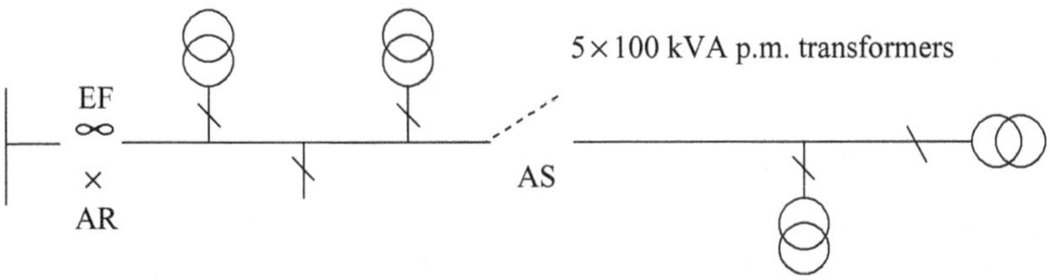

Figure 10.2   Rural network protection scheme

Table 10.1   Rural network protection costs

Scheme	Protection	Cost	Probable cons. h per annum	h per consumer
1	Expulsion fuses	£500	24 h × 500 cons. = 12 000	24
2	Autoreclose	£3500	4 h × 500 = 2000	4
3	Autosectionalise	£5500	1 × 300 + 4 × 200 = 1100	2.2

Note: probable consumer-hours (cons. h) interrupted per annum are calculated based on experience of annual interruption duration for such networks.

consumer-hours lost and an interruption of 24 h per consumer per annum (plus main network interruptions). An expenditure of £3500 will save 10 000 consumer-hours; an extra £2000 investment can still reduce interruptions by a further 1.8–2.2 h and consumer-hours lost by 900. The problem is which plan to choose? The estimation of interruption length is obtained through experience in operating such systems.

It is doubtful if any system planning will accept scheme 1 in Table 10.1 (except if there is acute capital shortage). The question is whether the extra cost of scheme 3 justifies the expenditure. In a mature supply the engineer will empirically decide that the 4 h are excessive and its reduction to 2.2 h justifies the expenditure.

Proper cost–benefit evaluation involves valuing the social cost of interruption to the consumers over the next 20–30 years, also the cost of the fuses repair, the AR and AS maintenance over the same period, and discounting these costs to their present value. If with scheme 2 the present value of the social cost of interruptions plus the discounted cost of maintenance is higher than £3500, then scheme 3 is chosen. Alternatively, in case of a lower cost, then scheme 2 is adopted. If the cost of interruptions is much higher, then a more reliable supply should be provided through an alternative source. This, of course, necessitates detailed evaluation of the cost to the consumer of aborted energy (curve (B) in Figure 10.1).

One interesting feature of this simple example is how the marginal cost per consumer-hour saved, increased considerably after making the first step, from £0.30 to £2.22. Thus, the dominant feature is the accelerating cost of higher reliability (curve (A) in Figure 10.1).

10.4   Detailed financial and economic evaluation

10.4.1   System adequacy and economics

The power system consists of three functional zones: generation facilities, transmission facilities and distribution. From the reliability point of view, it can be looked at in three hierarchical levels. The generation facilities, the generation and transmission and the third hierarchical level of the generation, transmission and distribution [1].

The adequacy of the generating system is the most important aspect in system planning to ensure the system-function performance. A shortage in the ability of generating facilities to meet demand can cause serious and sometimes widespread supply interruptions, particularly during peak demand. This leads to immense financial losses, inconvenience and disturbance to consumer welfare. Therefore, the adequacy of generation is paramount in power-system reliability. Adequacy of the bulk transmission network is also important. However, its reliability is usually much higher than that of other parts of the electrical power system. Although faults and inadequacies in this network can cause serious interruptions, they are less frequent than those of the other facilities of the system. Distribution system disturbances and inadequacy cause only localised interruptions, which are less serious than generation and transmission inadequacies.

10.4.2 Assessment and economics of generation adequacy

Sufficient reserve is essential to ensure the adequacy of the generating system. This reserve was determined, in the past, by methods such as ensuring that the generating system had enough percentage reserve, usually not less than 15–25 per cent of system peak, depending on the size of the system, in order to meet maintenance requirements, unscheduled breakdowns of generating facilities or higher demand than anticipated. In small systems, generation reserve was assessed to be equal to the largest or the two largest sets out of commission. Such deterministic methods cannot ensure the optimisation of investment in the system nor can they determine to a fair degree of accuracy the extent of system adequacy, and that this is the economically optimum adequacy. Correspondingly, simulation utilising probabilistic methods, which can more faithfully predict system performance and assist in the economical evaluation of investment, have become common in planning most large generating systems. The introduction of new generation facilities into the system reduces expected shortages and has system results that affect the whole economics of generation and the cost per unit of electricity produced. Such system effects can only be assessed through probabilistic simulation techniques.

The most widely utilised methods of assessing adequacy of generation are: the loss-of-load expectancy (LOLE) index, percentage energy loss index, and the frequency and duration method. These indices are used to assess probabilistic generation adequacy by convolution of the generating capacity model with the load model [8].

Loss-of-load expectation (LOLE) method

The LOLE may be the most widely used probabilistic index for generation reliability assessment. It is based on combining the probability of generation capacity states with the daily peak probability so as to assess the number of days during the year in which the generation system may be unable to meet the daily peak:

$$\text{LOLE} = \sum_L n_L A_G,$$

where n_L is the number of occurrences of peak demand L during the year, and A_G is the cumulative probability of being in capacity state j, such that $C_{j-1} \geq L > C_j$.

The reciprocal of the above equation is the LOLE in years per day. This means the probable number of years, or fraction of a year, during which the generation system will be unable to meet the peak demand of that day.

The LOLE index has gained recognition because it provides a probabilistic figure that can be computed and employed in system planning. It combines a generation and a load model. It is relatively simple to compute, understand and apply. However, this index suffers from the following shortcomings.

- It gives no indication of the extent of load shedding in mega watt (MW) or percentage wise, mega watt nor does it give any indication of its duration or frequency regarding the average consumer, which is what reliability assessment should be mainly concerned with.
- The LOLE, in days per year, mainly indicates the number of days in the year in which the generating system would not be able to meet the load. The frequency of load shedding may be higher than this figure in case of double peaked daily load curves and in systems that employ units with high failure rates but short repair duration.
- Since such an index is not capable of assessing the actual damage, it is not very useful for comparing the reliabilities of different utilities or national systems, particularly if they have different shapes of the load curve and peak durations.

The above arguments, particularly the first, have been recognised by the many users of this index. However, it is argued that for the same system, the use of the LOLE index would be adequate and correct for investigating different expansion plans and annual maintenance scheduling. This is only correct if the duration of peak demand is static over the years of study. This is not the case in many systems, with the continuous increase in the middle of the day load being experienced in most cases, particularly in developing countries.

To give an example, suppose that for a certain system with a typical daily load curve and annual load duration curves 'A' and 'a' respectively in Figure 10.3. The LOLE would be associated with the probably curtailment of the portion of the energy higher than the line 'C'. If over a few years, the shape of the load curve changes to that of 'B' and 'b' while the LOLE is maintained constant, then the probable amount of energy curtailment, area c + c'', increases faster than the growth of peak demand. This signifies a disutility to the consumer and an actual increase (probably small) in the unreliability of supply, which the LOLE cannot measure and defeats the purpose of maintaining the LOLE constant.

The effects of the change in the shape of the load curve and pattern of consumption, although they take place slowly over years, may change significantly from one season to another in the same year.

Besides the significant arguments against the LOLE mentioned above, it can be misleading in long-term planning and annual maintenance scheduling in systems of non-stationary load curves where significant changes in the duration of the peak load are expected to be encountered between seasons and over years.

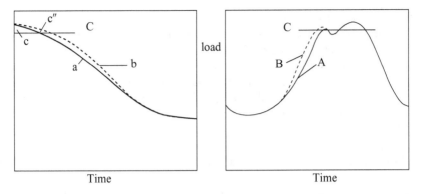

Figure 10.3   Effect of change of load curve on electricity curtailment

Frequency and duration method with a load model

The theory of the frequency and duration analysis of the generation states with a load model is detailed in Reference [8].

The generation frequency and duration approach, when coupled with the proper load model, will lead to the computation of a comprehensive reliability data. It will indicate the probable existence of all possible negative margin states, their frequency of occurrence, cycle time and mean duration.

However, the model does not yield a single reliability index. Two indices are calculated: an availability index and another frequency index. Here also the frequency index is of similar value to that of the LOLE and the sound arguments against its application for system expansion and medium-term operation can be applied. The availability index will account for the duration of load shedding but not its extent and magnitude.

This was recognised by the developers of this method and the following index of percentage energy loss (PEL) [9] was suggested as being more representative of the actual reliability of the system for planning purposes.

Percentage energy loss (PEL) index

The method involves the generation capacity states availability table and the daily load curve. The probable energy curtailment, during one year, divided by the energy requirement of the load, and multiplied by 100, yields the PEL index:

$$\text{PEL} = \frac{\text{probable energy curtailment}}{\text{energy requirement}} \times 100.$$

The probable energy shedding is obtained by combining the generation states availability table with the segments of the load curve. By summing up all load segments, the probable energy curtailment will be computed.

The percentage energy loss index comes nearer than any other single index to assessing the true reliability of generation, because it reflects the ratio of energy curtailment (disutility) to the consumer to his consumption (utility). Hence, it can measure the amount of inconvenience and loss to the consumer, which is the principal concern of any reliability criterion. However, it also suffers from the fact that it is not useful in a dynamic load curve. If the generation expansion scheme is aimed towards maintaining the index constant, then with the load factor improvement with time, owing to an increase in base load (say by an increase in off-peak space heating), the amount of energy permitted curtailment, which usually occurs during peak hours, would increase in proportion more than the growth in peak demand. This indicates deterioration in the 'actual' reliability of the supply and an increase in consumers disutility and inconvenience, which defeats the purpose of maintaining the index constant.

We have demonstrated the computational methods of various reliability indices that are widely utilised for power system projects planning. What is important is how to use them intelligently. This calls for a better assessment of the social costs of power system unreliability, particularly those caused by the generating system.

10.5   Financial and economic evaluation of quality of electrical power

The cost of electrical power interruptions [10] to facilities utilising electricity has been detailed earlier. However, these facilities can be sensitive to a wider range of power quality disturbances other than just outages that are counted in utility reliability statistics. In the new digital economy, momentary interruptions or voltage sags lasting less than 100 ms can have the same impact as an outage lasting many minutes.

To overcome the damage caused by these transient disturbances it is essential sometimes, for certain facilities (like banks, data and customer service centres), to invest in technologies for equipment protection and improving power quality. Such investments can be very high, and correspondingly a cost–benefit analysis is required before embarking on them.

Such a cost–benefit analysis will have to:

- analyse the power system quality performance,
- correspondingly estimate costs associated with power disturbances,
- look into technological solutions in terms of effectiveness and cost, and
- perform the cost–benefit analysis.

Voltage sags and monetary interruptions have most important impacts on performance of facilities. There is equipment that is usually sensitive only to magnitude of voltage variation, others are sensitive to both magnitude and duration of an r.m.s. variation.

IEEE Standard 1159 defines voltage sags lasting between 0.5 and 30 cycles as 'instantaneous', those lasting between 30 cycles and 3 s are identified as 'momentary', and those lasting between 3 s and 1 min are defined as 'temporary'. In addition to magnitude and duration, it is often important to identify the number of phases involved in the sag.

The costs associated with sag events vary widely according to facilities and market conditions. Such costs can be product-related losses, labour-related losses and other costs like damaged equipment and loss of income. Costs will vary with magnitude, duration and frequency of power quality disturbance.

Cost of power quality improvement technologies depends on the alternative category involved, its investment cost as well as operating and maintenance cost. For small control protection (less than 5 kVA) CUTS, UPS, and dynamic sag protectors are employed. For machine protection (up to 300 kVA): UPS, flywheel as well as dynamic sag corrector are employed. For large facility protection in the 2–10 MVA range, beside the UPS and flywheels, dynamic voltage regulators as well as static and dynamic transfer switches are employed.

To perform the required economic analysis the solution effectiveness of each alternative must be quantified in terms of the performance improvement that can be achieved. Solution effectiveness, like power quality costs, typically will vary with the severity of the power quality disturbance, and the type of activity.

Once the costs of the quality improvement technologies and their effectiveness have been quantified, a normal cost–benefit analysis exercise, like the ones explained in this book, have to be undertaken before decision making.

10.6   Evaluation of investment in generation

Optimisation of investment in the generating system is important for the economics of the electricity supply industry. Generation facilities constitute almost two-thirds of all new investments in the electric power system and the vast majority of the operational cost. The losses due to their unavailability are widespread in contrast to the network problems, which have localised consequences. Therefore it is essential to optimise these to ensure the economic as well as the technical adequacy of the power system.

The most important points to be addressed when investing in power generation are as follows.

- *Selecting the location* – many technical and economical considerations govern selecting the location (fuel availability, cooling water requirements, availability of land, proximity to the load centre, configuration of the transmission network and environmental considerations).
- *Timing* – this is influenced by the criteria for generation adequacy as well as load growth prospects and lead time for the new facilities. Such issues were addressed in Chapters 5, 6 and 7.
- *Type, size of the new facilities and fuels* – there are many and increasing types of new generation facilities (conventional thermal units firing pulverised coal or other fuels, nuclear, hydro, gas-turbines in a simple or combined cycle, etc.). What is important is not only the type of facility and fuel consumed, but also the optimum set size that fits economically, to the growing load curve, in order to reduce risk.

There are several established algorithms to assist in generation planning and its economics. All of these are based on a decision-making framework expressed in economic terms aiming to schedule least-cost investments for power generation expansion. Some of these are Wein Automatic System Planning (WASP) Model [11] and the UNIPEDE Model [12]. Such models, however, fall short of achieving proper financial and economic evaluation of generating planning because of weaknesses in the system adequacy criteria that they utilise, as explained above. It is necessary to undertake detailed financial and economical evaluation, with proper economic adequacy criteria.

10.6.1   Valuing cost of interruptions

The cost of electricity to a consumer, i.e. the consumer's evaluation of the worth of the electricity supply (while ignoring consumers surplus), is equal to payments for electricity consumed plus the economic (social) cost of interruptions.

Supply interruptions cause disutility and inconvenience, in varying degrees and in different ways, to different consumer classes, domestic, commercial and industrial. The costs and losses (L) of these interruptions to the average consumer are a function of the following:

- dependence of the consumer on the supply (C)
- duration of the interruption (D)
- frequency of its occurrence in the year (F)
- time of the day in which it occurs (T)

i.e.

$$L = (D^d \times F^f, T^t) \times C,$$

where d, f and t are constants, but vary from one consumer category to another.

It is becoming increasingly necessary to assess the cost of interruptions to consumers in monetary terms and to use this assessment as an input in the project evaluation to arrive at the system plans that minimise total overall financial and economic cost to the utility and the consumer (payment in tariffs plus the cost of interruptions). Many problems are encountered.

- Consumer cost is not only a function of the frequency and duration of the interruptions, but also the day and the time of the day in which they occur as well as consumers category and expectations. An interruption that occurs after midnight and causes a considerable amount of energy curtailment of off-peak heating may pass unnoticed by the consumer, while a similar interruption that involves much less energy during peak hours and household maximum activity will cause enormous consumer irritation and inconvenience.
- Consumer costs due to interruption are not a linear function. It is highly likely that they will be similar to curve (B) in Figure 10.1, i.e. the marginal utility of an extra incremental improvement in reliability decreases slightly as reliability improves.

- Generation shortages usually occur during peak hours where system demand is highest and value of electricity is most important to consumers. Most network-caused interruptions are random in occurrence and can happen any time during the day. Therefore interruptions caused by generation shortages are more costly to consumers than network problems.
- The amount of energy curtailed is no indication of the value of electricity to consumers. The value of operating little-energy consuming sensitive machinery for a firm, or TV and light to a household, far out-weights the value of high-energy consuming apparatus and machinery, at least for short interruptions.

Until now and in most countries, the determination of the level of reliability is based on experience, judgement and also the availability of funds to invest in standby plant and redundant systems. Increasingly in Europe and North America, a lot of techniques are used to assess the financial losses to consumers as a result of supply interruptions, i.e. the worth of reliability in monetary terms. There are many ways to assess such costs; most of these can be obtained from the established techniques of *consumer surveys*. The purpose of consumer surveys is to assess consumer *willingness to pay* (WTP) for extra reliability of supply. From consumer surveys it is possible to draw out consumer *damage functions* (consumer cost). These reflect the costs that consumers of different categories are likely to endure as a consequence of supply interruption, based on their demand, category of use, and the frequency and duration of these interruptions. Such costs are weighted with regard to the respective energy utilisation in the district (or country). The weighted costs are then summed to provide the total cost for the district for each specified duration. These are termed as *composite consumer cost function* for the district or country. From these the total annual cost of interruptions is computed and compared with the energy curtailed to arrive at an average economic cost per kWh of electrical energy curtailed [13,14].

There are some uncertainties that surround such an approach, but it is closer than anything else in assessing the actual cost to consumers of supply unreliability. This method can be refined further to assess a rate for generation interruptions and a different rate for network interruptions, taking into account that most generation shortages occur during peak hours. Therefore, energy curtailment at peak times is more valuable than that resulting from transmission and distribution network problems, which are random in nature. The total value of the electricity to a consumer is the amount of the payments made by the consumer, in the form of tariffs, plus the social cost of electricity interruptions, as valued by the consumers' willingness to pay.

Economic cost of electricity = (energy consumed in kWh × tariff) + social cost of energy interruptions (kWh interrupted × average cost to consumer per kWh curtailed).

10.6.2 Incorporating reliability worth in generation planning

The economic planning of electricity production will try to minimise this economic cost by striking the right balance between electricity production cost (tariff) and supply continuity, as detailed in Figure 10.4.

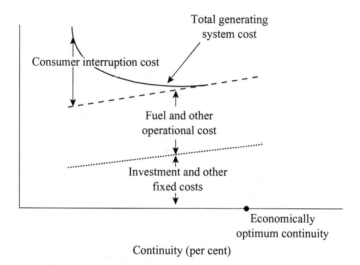

Figure 10.4   Total cost of the power system including cost of interruptions

Such a calculation will indicate the percentage continuity figure that minimises total generation system cost (including interruptions cost), and correspondingly the amount of percentage reserves required to attain such a continuity.

This cost factor is useful for power system planning in order to minimise the system cost to consumers. In such cases, a generation simulation model like that of Figure 10.5 is utilised to assess the expected (cost of generation at a future year [cost of electricity (fixed and operational) + cost of interruptions to the consumers). Sets are added until the optimum continuity is obtained. This will occur when the above sum is at a minimum. Such simulation is carried out for many years into the future, utilising different generation-extension scenarios, in order to plan a programme for generation strengthening over time.

The same approach applies to the planning and the timing of network strengthening, where the economic cost of electricity supply to a particular area is equal to the cost of energy utilised by the area plus the cost of interruption to the consumers in that area, as caused by network problems. A network-strengthening exercise is carried out to assess the economic cost of each new network-strengthening configuration in order to compute the discounted net benefits of each scheme. The net benefits in this case will be the discounted reduction in consumer economic cost due to interruptions plus reduction in losses minus the discounted cost of the network-strengthening scheme (including any other system cost). If these net benefits are positive, then the strengthening scheme is undertaken. It has to be recalled that the discounted reduction in consumer economic cost is equal to the discounted amount of reduced energy curtailment in kWhs multiplied by the average economic (social) cost of each kWh curtailed.

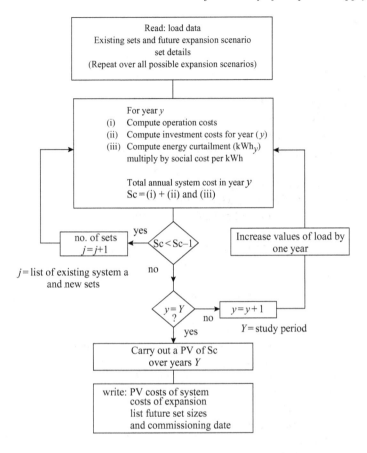

Figure 10.5   System expansion plan (to minimise total system cost (operation + investment + interruptions))

Many attempts were made to assess the economic cost of each kWh curtailed in order to utilise this in system planning. In the 1990s, the UK electricity supply industry assessed the value of lost load (VOLL) to be £2.345 kWh^{-1} [15]. The VOLL is defined as the value that customers place on the amount of energy they would have consumed during a supply interruption. A study [13] in North America in 1985, based on the computation of consumers damage function, assessed this to be equal to almost $5 kWh^{-1} (£3.00), which is not very different. However, such values are much lower for developing countries, where electricity is valued by the consumer at a lower value owing to their limited WTP, and the shortage of funds, particularly investment prospects in foreign currency, necessities that lower levels of reliability be tolerated.

It has to be realised that such values are based on certain supply continuity standards and consumer expectations. These are likely to change with changes in continuity as demonstrated by curve (B) in Figure 10.1.

10.7 References

1 BILLINGTON, R., and ALLAN R. N.: 'Power-system reliability in prospective' (Applied Reliability Assessment in Electric Power Systems, IEEE Press, 1991, pp. 1–6)

2 BILLINGTON, R., and ALLAN, R. N.: 'Power system reliability in perspective', *Electron. Power, J. Inst. Electr. Eng.*, March 1984, pp. 231–236

3 KHATIB, H., and MUNASINGHE, M.: 'Electricity, the Environment and Sustainable World Development', World Energy Council, 15th Congress, Madrid, 1992

4 'Power Shortages in Developing Counties' US Agency for International Development, report to Congress, March 1988

5 KHATIB, H.: 'Economics of reliability in electrical power systems', (Technicopy Ltd., England, 1978)

6 MUNASINGHE, M.: 'The economics of power system reliability and planning' (World Bank Research Publication, 1979)

7 BILLINGTON, R., WACKER, J., and WOJCZYNSKI, E.: 'Comprehensive bibliography on electrical service interruption costs', *IEEE Trans.*, June 1983, **PAS-102**, (6), pp. 1831–1837

8 IEEE Committee: 'Bibliography of the application of probability methods in power system reliably evaluation', *IEEE Trans.*, 1984, **PAS-103**, pp. 275–282

9 BILLINTON, R., WACKER, G., and WOJCZYNSKI, E.: 'Customer damage resulting from electric service interruptions', Canadian Electrical Association, R&D project 907, U131 report, 1982

10 MCGRANAGHAN, M.: 'Economic Evaluation of Power Quality', *IEEE PER*, February 2002, **22**, (2), pp. 8–12

11 'Senior Export symposium on Electricity and the environment', Helsinki, May 1991 (International Atomic Energy Agency (IAEA), Vienna 1991)

12 'Electricity generating costs for plants to be commissioned on 2000', UNIPEDE, Paris, January 1994

13 BILLINGTON, R., and OTENG-ADJEI, J.: 'Cost/benefit approach to establish optimum adequacy level for generating system planning', *IEE Proc. C*, March 1986, **135**, (2)

14 BILLINGTON, R., OTENG-ADJEL, J., and GHAJAR, R.: 'Comparison of two methods to establish an interrupted energy assessment rate', IEEE *Trans.*, August 1987, **PWRS-2**, (3)

15 ALLAN, R. N., 'VOLL-fact or fiction?', *Power Eng. J.*, February 1995, **9**, (1), p. 2

Chapter 11

Electricity trading

11.1 Introduction

The energy sector is undergoing a major transition worldwide [1,2]: competition, restructuring, privatisation and regulation are at the core of the current revolution, driven in part by new technological developments and changing attitudes towards utilities. The objectives of these reforms are to enhance efficiency, to foster competition in order to lower costs, to increase customer choice, to mobilise private investment, and to consolidate public finances. The mutually reinforcing policy instruments to achieve these objectives are the introduction of competition (supported by regulation) and the encouragement of private participation. An international approach for the design of the legal, regulatory, and institutional sector framework has emerged. It includes the following.

- The corporatisation and restructuring of state-owned energy utilities.
- The separation of regulatory and operational functions, the creation of a coherent regulatory framework, and the establishment of an independent regulator to protect consumer interests and promote competition.
- The vertical unbundling of the electricity industry into generation, transmission, distribution, and trade (services).
- The introduction of competition in generation and trade and the regulation of monopolistic activities in transmission and distribution.
- The promotion of private participation in investment and management through privatisation, concessions, and new entry.
- The reduction of subsidies and rebalancing of tariffs in order to bring prices in line with costs and to reduce market distortions.

One of the most significant features of the electricity supply industry in recent years has been the emergence of electricity trading in liberalised electricity markets; this also fostered risk management practices. Such activities, particularly risk management, are unimportant in regulated markets with fixed prices; it is only in liberalised markets where prices are charged in accordance with supply and demand and future expectations with possible price volatility that such activities flourish. This is enhanced by the introduction of customer choice. Such features manifest themselves to a varying degree in various liberalised markets; however, they are prominent in the US where prospects for electricity price volatility exist, owing to supply and delivery restrictions, more than in the UK and Europe where there are abundant reserve margins and strong interconnected transmission networks. Market players include generators producing electricity, suppliers who buy electricity to sell on to groups of consumers, traders and marketers who do not own generating assets but have active roles in the market place, and other players who provide, for example, risk management, hedging and brokerage.

Power marketing (or trading) refers to any number of financial and/or physical transactions associated with the ultimate delivery of a host of desirable energy-related services and products to wholesale and, increasingly, retail customers. Power marketers, those engaged in such trade, however, need not own any generation, transmission or distribution facilities or assets. They rely on others for the physical delivery of the underlying services. Moreover, power marketers operate primarily as contractual intermediaries, usually between one or more generators and one or more customers.

The physical nature of electricity does not allow a true spot market (instant pricing and delivery) so financial transactions must be scheduled some time in advance of physical delivery, with pools or power exchanges thus substituting for a true spot market.

Electricity market trading is quite different from commodities trading (or other forms of energy trading) because of the nature of electricity – it cannot be stored, its availability must be instantaneous and absolute, as well as the technical complexities of the expertise, knowledge and planning capabilities that only power engineers can provide. For an electricity market to perform successfully, two types of expertise must converge:

- a high level of technological expertise in the domain of power engineering, and
- financial and business expertise allowing market trading.

Only after the power engineers have set the parameters for physical delivery, both short term and into the future to meet foreseeable conditions, can the financial market trading element of the electricity markets come into play.

11.2   Electricity trading worldwide

For regulators, the creation of trading exchanges can offer the chance to build a truly open and competitive market, guided by a global knowledge base of the successes and failures of other exchanges in other industries around the world [3]. It is interesting that the Financial Services Authority (FSA) will regulate energy trading in the UK, rather than the industry regulator (OFGEM), a trend that will continue in Europe.

Energy exchanges enable the development of the wholesale business. In addition to the trading of physical quantities, 'future' markets are created making extensive use of financial products. Many exchanges offer multi-energy (i.e. electricity, gas, and oil) services, sometimes extending to other commodities as diverse as metal, pulp, and paper.

The number and nature of players will evolve as the electricity market continues to open and the liquidity of exchanges increases. It is foreseeable that electricity trading will occur increasingly over the Internet in the coming years. Some projections anticipate that, by the year 2004, approximately 11 per cent of total electricity and about 25 per cent of natural gas sold in the US will be sold online.

There is a lot to be gained for all parties through these new markets. But it can be a complex process, and companies should evaluate participation in a trading exchange against the current market trends, the drivers in energy markets and the broader developments in financial and commodity trading.

Such considerations are unlikely to lessen the pace at which trading exchanges in the energy sector are growing. Instead, market forces, technology, and legislation will shape the new exchange landscape, creating an environment in which competition increases rapidly and consolidation occurs. It is vital that this moulding influence is allowed to continue, as for a market to successfully move to a deregulated mode, the basics such as maintaining an adequate balance of regional supply and demand must be established.

Across the world, competition in energy markets has driven the development of wholesale energy trading. There is an enormous variety in the speed and willingness of markets to deregulate, from country to country, and even from state to state. Many countries already have fully competitive and mature markets while other countries still do not plan to deregulate their gas and electricity industries.

The US is the most important market for electricity trade; this market and its development will be explained in Section 11.2.1, and then the UK and other markets will be discussed in Sections 11.2.2–11.2.5.

11.2.1   Electricity trading in the US

During the past decade or so, and particularly since 1992, the growth of electricity marketing, particularly in the US has been phenomenal [1]. The year 1978 saw the passage of the Public Utilities Regulatory Policy Act (PURPA), which opened the power-generation sector to new players, mostly independent power producers (IPPs). For IPPs to prosper and succeed they needed access to the national transmission network in order to deliver their product to large consumers. This was provided for in the Energy Policy Act of 1992, and was supported by two significant orders, in 1996, in which the US Federal Energy Regulatory Commission (FERC) spelt out the modalities for implementations on how an open access transmission system would work in practice. More specific orders were issued in 1999 and 2001 encouraging and calling for the creation of regional transmission organisations (RTOs). Such major milestones in the development of the liberalised US electricity markets are summarised in Table 11.1.

In North America, the electricity market is deregulating at a variable pace. In the US, each state moves forward independently from the others. California, which is further along the process, has experienced significant difficulties (see Section 11.2.4)

Table 11.1   Electricity milestones: major laws with significant impact on US electricity markets

---

1935 Federal Power Act
Created the FERC and established principles for regulating wholesale electricity pricing.

1978 Public Utility, Regulatory Policy Act (PRUPA)
Allowed independent power producers (IPPs) to flourish.

1992 Energy Policy Act (EP Act)
Introduced the premise of a non-discriminatory open access transmission network.

1996 FERC Orders 888 and 889
Spelt out FERC's long-standing policy on how an open access transmission system would work in practice; Order 889 spelt out the details of the Open Access Same-Time Information System (OASIS).

1999 FERC Order 2000
Encourages the establishment of regional transmission organisations (RTOs).

Source: Energy Informer (July 2001), Reference [1].

that have influenced the pace of deregulation in other states. For example, states such as Oregon, Nevada and Arizona have put the decision to deregulate on hold. Some Canadian provinces are taking a comparable path to deregulation.

11.2.2   UK electricity market trading

In England and Wales [2] before privatisation began, the electricity industry was a classic, vertically integrated, government-owned monopoly, seen at that time as the best way to provide a secure electricity supply. Consumers had no choice of supplier and had to buy electricity from their local regional electricity company (REC), so that price competition was not possible.

The UK is one of the pioneer countries in developing a free market electricity trading system. On 27th March 2001, New Electricity Trading Arrangements (NETA) for England and Wales were launched. The stated objectives of NETA are to benefit electricity consumers through lower electricity prices resulting from the efficiency of market economics. Promotion of competition in power generation and electricity supply, in order to use market forces to drive consumer costs down, was, and remains, a key objective of actions to liberalise and 'deregulate' electricity markets in the UK.

In England and Wales, the electricity market was progressively opened to competition in generation and supply over a period of 10 years, with retail competition for the smallest consumers completing full market liberalisation in 1998. A regulatory body, the Office of Gas and Electricity Markets (OFGEM), oversees market operations, licensing, competition, and actively seeks to make changes to benefit consumers and bring their costs down while promoting competition between players in the electricity market. Sweeping away old thinking has led to electricity being regarded as a tradable commodity and revolutionised the business environment bringing new market participants, new business opportunities and rapid change.

Initially market reform involved creating an Electricity Pool for England and Wales with a single wholesale electricity price. Producers sold to the Pool and licensed suppliers purchased electricity from the Pool. Pool participants were able to negotiate bilateral contracts. However, the Pool performances did not allow the development of full competition. The UK White Paper on Energy Policy and the Utilities Act 2000 incorporated proposals for trading arrangements similar to those in commodity markets and energy markets elsewhere. NETA provided new structure and rates for England and Wales electricity market. Under NETA there were major developments in which electricity is bought and sold, with major competition in generation and supply, with a wide range of new players competing in the liberalised energy market as detailed in Figure 11.1.

NETA aims to develop a marketplace in which buyers and sellers can participate through a variety of contracts and agreements within the framework set by NETA. The transactions taking place within the NETA market are electricity price–quantity transactions on a half-hourly basis.

Market players include generators producing electricity, suppliers who buy electricity to sell on to groups of consumers, traders and marketers who do not own

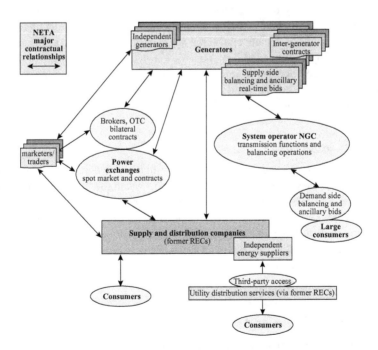

Figure 11.1   NETA electricity market – major contractual relationships (Reference [2], OFGEM)

generating assets but have active roles in the market place, together with other players who provide, for example, risk management, hedging and brokerage.

NETA is a new wholesale market, comprising trading between generators and suppliers of electricity in England and Wales. Under NETA, bulk electricity is traded forward through bilateral contracts and on one or more power exchanges. NETA also provides central balancing mechanisms, which do two things: they help the National Grid Company (NGC) to ensure that demand meets supply, second by second; and they sort out who owes what to whom for any surpluses or shortfalls. The majority of trading (98 per cent in the first year) takes place in the forward contracts markets. A very small percentage of electricity traded (2 per cent in the first year) is subject to the balancing arrangements.

It is hoped that most of the trading will be through bilateral contracts – forwards and future, which spot market trading utilised to adjust contractual positions. The nature of the ESI does not allow spot pricing so that forward arrangements of the financial transactions are necessary.

Under NETA the market provided through power exchanges replaces the previous Pool arrangement, allowing market players to trade electricity up to one day ahead of the requirement for physical delivery. The National Grid Company (NGC) operates as a system operator for England and Wales, managing the HV transmission system and also providing all the technical and operational services normally demanded by the system to ensure its integrity including load forecasting, ensuring system security and stability, frequency control, and reactive power control. NGC acts on both a physical and a financial level through the balancing mechanism, selecting bids and offers for incremental or decremental supply of electricity in order to achieve physical balance between generation and demand.

It is not intended here to go into the way the financial and settlement activities are undertaken in the England and Wales electricity market trading since these are explained in detail in the literature [2].

11.2.3   The Nordic market

This involves Norway, Sweden, Finland and Denmark. The deregulation of the electricity market in these countries is complete with customers free to choose their supplies. Trade of electricity (mainly generated by hydro-power) between the four countries is extensive. In this market, generation and retail sales are competitive while transmission and distribution are regulated. Third-party access to the transmission network is regulated; therefore there are equal access rights to all users. The Nordic network experience provides an excellent experience to European Electricity market development. The Nordic electricity market general relationships are in Figure 11.2.

11.2.4   The California market

The pioneer California market provided the most severe challenge to competitive electricity market philosophy. Restructuring of the ESI of California took place in 1996, with the aim of bringing the benefits of competition to consumers. Prior to this

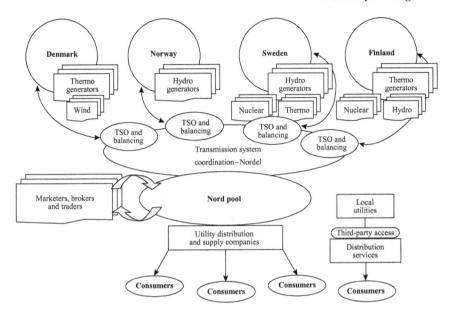

Figure 11.2   The Nordic market – major contractual relationship [2]

regional utility companies – investor-owned utilities (IOUs) – provided monopoly supply and services. These former utilities now each provide a regulated distribution service in their areas, allowing direct access to third-party energy-service suppliers; consumers now have a choice of electricity supplier.

When the new California electricity market structure took effect, the utilities had the prices for their consumers frozen at 10 per cent below the level at vesting in the expectation that costs and prices would fall. It was anticipated that consumer electricity prices set at this level would allow the utilities to recover the cost of investments that had been made before market liberalisation stranded costs. As events unfolded, this proved to be entirely unfounded and resulted in the utilities business becoming non-viable.

The crisis in the California electricity markets resulted from a combination of factors [2].

- Exceptionally high summer temperatures significantly increasing electricity peak demand.
- A lack of generating capacity in California and the West of America in relation to the strong growth of electrical demand following economic growth in California.
- A shortage of water resulting in relatively limited hydro-power import availability from the North West of America.
- An increase in gas prices for power generation compared with previous years.
- Exercise of market power by generators and other market players.
- Environmental restraints on the construction of new generating plant and operation of existing plant.

- Weakness and flaws in the design of the electricity market including limitations on forward contracting, fixed consumer prices, but variable wholesale electricity prices.
- Insufficient importance being given to power engineering expertise in design of market structures.

Shortage of installed capacity or plant availability due to outages, maintenance, or generator market power, seems to have been a key driver for the California difficulties. Structural weaknesses in the design of the California market include restraints on consumer prices but free competition in wholesale electricity prices and constraints on utilities to buy through a power exchange. These factors were major contributors to the problems of the California electricity market. In order to ensure proper and reliable market trading it is imperative to ensure a technically viable and reliable system. Electricity markets demand technological expertise in power engineering plus a financial and business expertise that allows market trading.

What California tried was not really deregulation. What failed were price controls on retail electric rates, and an overly restricted, supposedly deregulated wholesale market – a mix that was definitely asking for trouble. California's experience has been a bit of a horror story, and truly is not a valid case against deregulation. It is, instead, a case study of how not to 'deregulate'.

11.2.5 Electricity markets across Europe and elsewhere

The European Commission Single Market Directive for Electricity came into effect in February 1999. The Directive obliges EU Member States gradually to open their power sectors to competition; to vertically unbundle the sector; and to ensure nondiscriminatory access to the transmission network.

Energy markets across Europe are expected to be fully liberalised, privatised, and integrated across borders by 2010. European companies and households will benefit from lower prices, better services, and free choice between alternative providers. European utilities will be highly competitive as a result of cost cutting and consolidation. The EU Single Market for energy is expected to be the largest in the world, comprising not only the current 15 Member States, but also up to 13 accession countries, with a total of more than 400 million consumers.

Germany leads the European countries in progress towards deregulation and is the region's largest electricity market by consumption. It is strategically important as a base for further European expansion, with physical interconnection to many more European markets than the UK. The country has two regulated exchanges, which have merged into one entity. However, there is little liquidity on this market and brokers have been more successful in trading, most notably through electronic trading in the form of Alternative Trading Systems (ATS).

Australia and New Zealand were among the first deregulated countries, and have almost completed their transition and are refining and adjusting the implementation of the new market rules. Singapore and Malaysia were expected to experience a full opening of their markets in the third quarter of 2002, whereas Japan is only now seeing a partial opening to competition in the construction of merchant power plants and the electricity supply to large customers.

North America pioneered reforms in the 1980s, but owing to its federal structure has not yet completed the process in all states. Except for the UK and the Nordic countries, Europe has embraced reforms relatively late but vigorously, and is now arguably the fastest reforming continent. Latin America, the first developing region to liberalise and privatise its energy sector, has largely completed the reform agenda. Many countries in Asia that introduced IPPs without liberalisation suffered from the consequences during the recent financial crisis, and are now moving towards the Latin American model. Together with Sub-Saharan Africa, the countries of the southern Mediterranean are lagging considerably behind international reform trends [3].

At the beginning of 2002, there were approximately ten regulated electricity exchanges in Europe, with eight ATS trading electricity in the region and at least three more regulated exchanges and a further three ATSs close to introduction. With wholesale markets at very different stages of development in each country it is only fair to describe the European picture as fragmented and largely immature, characterised by the number of regions and exchanges; the lack of appropriate instruments; issues around transparency and liquidity; and restricted inter-regional trading.

But the immaturity of electricity trading in most countries has not prevented the creation of new trading exchanges. Indeed, in some new territories, particularly those in which legislation supports competition, an under-developed market offers greater potential gains once maturity and liquidity is reached.

The UK is a good example of this. At the beginning of 2002 there were four exchanges competing for market share in a region where electricity consumption is below the European average. The fight is on between these exchanges, partly because the region presents a considerable reward, but also because it will act as a good proving ground, offering a chance for exchanges to cut their teeth before moving on to markets that are less developed.

Generators, wholesalers and retailers access exchanges through the trading activities of internal business units or by outsourcing to trading companies. As electricity prices are extremely volatile, successful trading activity depends on a good risk management policy and implementation. The number and nature of players will evolve, as the electricity market continues to open and the liquidity of exchanges increases. The increase in electronic trading of all forms has resulted in the need for companies to reassess the way in which they trade and the way in which electronic trading impacts on their risk policy.

11.3   Electricity traders

Because of price volatility in the bulk supply markets the importance of electricity traders has increased during recent years [1]. Correspondingly risk management and hedging against such volatility is one of the critical services offered by electricity traders. This is likely to be extended soon from the bulk supply market to the retail competitive markets. Electricity traders have access to valuable information (prices, options, etc.), which will also assist their clients in reducing costs, increasing profits and improving performance. They can also broker and undertake transaction management and assist in providing liquidity by helping their customer to change position quickly.

The electricity-related services provided by power traders include physical delivery of power and/or a financial obligation or promise to do so. Such services extend beyond electricity into other energy-related services like natural gas. The delivery may be firm or non-firm, short or long term, one time or stretching over time. Prices may be firm or indexed to other commodities or derived from combination of underlying commodity prices, hence 'derived markets'. Many operators in the electricity business will not be able to function successfully in the liberalised competitive electricity market without the help of an electricity marketer.

Electricity marketing has to investigate customer needs, decide on the prices to offer and the considerable risks and how to manage them through mitigation and hedging strategies. Small, uncorrelated risks may simply be managed through risk pooling. Aggregating a large number of small risks, which are not correlated, is the easiest way to manage risk (see Chapter 14).

As from the beginning of the 21st century, electricity trading is growing at a phenomenal rate. In the US the annual volume of wholesale electricity commodity trading was anticipated to reach $2.5 trillion by the year 2003 and online transactions to reach $4 trillion by 2005 (see Figure 11.3). Such amounts will make electricity the single largest traded commodity in the US. By 2003 the $2.5 trillion was expected to be about 10 times the retail value of electricity trade in the US. This is made possible by the developments in financial trading instruments such as futures, forwards, swaps, and options that allow the same electrons to be bought and sold several times before they are actually generated and consumed. The volume of trade is typically many

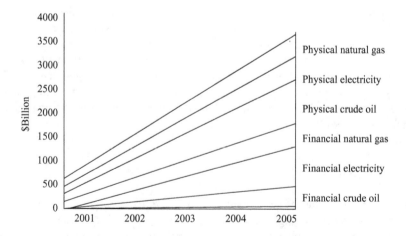

Figure 11.3   Online   energy   trading   projections   (Forrester   Research   and Reference [1])

times the physical number of kWh generated or consumed because many times an electricity trader will be buying from or selling to another trader. Correspondingly the same kWh is usually bought and sold a number of times until it is finally delivered and consumed.

The services an electricity trader can offer are discussed in Section 11.3.1. The most important is the quantification, management and hedging of risk.

11.3.1 Case study: sample of services offered by power marketers (traders)

The services most commonly offered are categorised as follows [1].

Facilities management

Many utilities and customers can benefit from improvements in the utilisation, maintenance and operation of their existing assets, plants, infrastructure and personnel. This is particularly true for some of the smaller utilities that may not have adequate in-house or internal resources for such services, and outsource them to others. These services are usually provided in addition to more traditional power-marketing functions such as selling the excess output of the plant (e.g. what is not needed to serve the utility's native load) to the highest bidders outside the utility's immediate service area.

The savings resulting from the improved operations and maintenance (O&M) practices are usually shared among the two parties under pre-arranged terms. The power marketer's willingness to manage certain critical O&M functions, assume risks, and arrange financing makes such schemes highly attractive to many clients who do not have the internal resources or expertise to engage in such functions. Since no up-front investment may be needed, and no risk is borne if there are no improvements in performance of the plants, there is little risk to engage in such deals with power marketers.

Risk management

Offering a variety of services to provide price stability and risk hedging is among the most sought-after services provided by power marketers. Since many customers and smaller utilities are not familiar with, and do not have the necessary skills and capabilities to manage, the risks internally, they rely on power marketers for such services.

Total energy services, solutions, maintenance, etc.

By combining many types of input fuels (e.g. natural gas, coal, oil) and end-use energy and/or services (e.g. electricity, natural gas, steam, hot water, and others) it may be possible to reduce a customer's overall energy costs while at the same time providing an improved level of service. For example, by switching from natural gas to coal on the fuel side, and/or by switching from electricity to natural gas on the end-use side, a customer can take advantage of lower price opportunities.

Tolling services

The power marketers' ability and willingness to buy raw energy at low prices in one location or time and to deliver a portfolio of desirable energy services and customised products at the another location or time is highly attractive. Although customers can, in principle, perform these functions on their own, many do not have the market reach or the risk-management capabilities of the power marketers. Moreover, power marketers, by virtue of having many offsetting tolling transactions, can balance the risks of market price fluctuations through pooling and other arrangements, thus reducing the risks to the individual customers.

Structured/customised products

Structured or customised products are increasingly emerging to meet the specific needs of sophisticated customers. Another reason for their increased popularity among power marketers is that customers are willing to pay significant premium for these types of services. Offering a package of customised goods and financial services that meet specific customer needs, their financial resources, and risk tolerance, however, tends to be time- and resource-intensive, hence the higher premiums.

These types of services cannot, by their nature, be standardised. They require extensive effort to negotiate and complete. But the results could be worth the extra effort.

11.4   Future strategies and growth

Although the introduction of competition in electricity markets is high up the European agenda, there is still some way to go before market integration is complete [4]. As a result, exchanges need to adopt country-by-country strategies, recognising the deregulation agenda and conforming to the competitive environment of each individual country. In essence, exchanges need to adopt a portfolio approach that encompasses corporate strategy, the market environment of each country and the competitive landscape. The legislative environment and regulatory regime of each country need to be considered, as well as the differences in the country's market potential, market participants and competitors. Potential partnerships and alliances can be vital and these opportunities should not be overlooked.

The surge in the development and growth of electricity exchanges around the world has largely been driven by the triumvirate of market forces: legislation, competition and technology. This trend is likely to continue in the short term as legislation in Europe and North America increases the number of new markets opening. In the medium and long term, however, customer demand and increasing competition will be the driving forces.

The world's electricity markets are fragmented and immature; and the European electricity industry, for all its successes, is typical of the global situation. While there are already many exchanges, both of the regulated exchange and the ATS varieties, there will be continued growth in the number of exchanges in the short term.

Much of this growth in numbers could be reversed in five years' time, as a strong period of consolidation results in the birth of mega-exchanges. During this time, there will be unprecedented access to new markets through deregulation, better integration of the fragmented markets in Europe and North America, and truly mature markets will develop in these regions, both in electricity and gas. This represents significant change and unprecedented opportunity for energy traders and exchanges alike. Being successful requires the right strategy, the right partnerships and the ability to execute those plans globally and on a portfolio basis.

11.5 References

1 SIOSHANSI, F. P.: 'Electricity trading', *Energy Policy*, 2002, **30**, pp. 449–459
2 STEPHENSON, P., and PAUN, M.: 'Electricity market trading', *Power Engineering Journal*, 2002, pp. 277–288
3 MULLER-JENTSCH, D.: 'The Development of Electricity Markets in the Euro Mediterranean Area', *The World Bank Technical*, Paper No. 491, 2001, Washington D.C
4 COOPER, C.: 'Survival of the fittest', *Electricity International*, February 2002, **14**, pp. 14–20

Chapter 12

Evolvement of the electricity sector – utility for the future

12.1 Introduction

During the last decade of the 20th century the electricity supply industry underwent major evolvement and restructuring [1]. Until very recently most network industries (mainly the electricity supply industry, but also gas, water and telecommunications) were considered to be natural monopolies. Their size, networks, capital-intensive nature and their sensitive services to the public meant, in most cases, exclusive government ownership and control. Recently, however, technological progress, particularly in information technologies and telecommunications as well as development of regulatory instruments has enabled the introduction of a market mechanism into these traditional monopolies. Developments vary from one country, or region, to another but a general pattern of four phases developed.

12.1.1 Phase I: early 20th century – private sector investment and monopolistic market behaviour

The infrastructure investments in the late 19th and early 20th century were largely undertaken by private companies. Private firms developed and commercialised the technologies for the production and delivery of electricity and natural gas. Local monopolies, and national and international oligopolies that used their market power to extract economic rents from captive customers, dominated the new industry. Delivery to users was generally confined to urban communities, with limited development of distribution grids in rural areas. There was little competition in the sector during this period of rapid innovation and industry expansion.

12.1.2 Phase II: mid-20th century – public sector intervention and inefficiency

Around the time of World War II, a trend towards the nationalisation of energy assets or at least strong government regulation of privately owned monopolies became the norm, in an attempt to limit abuses of market power. In many countries, governments also played an important role in rural electrification, since returns were too low to attract private capital. Elsewhere, state ownership of the electricity industry became the rule. Over time, however, public ownership and the absence of competition increasingly undermined effective management, innovation and operational efficiency. Governments used the power sector, like other state-owned industries, artificially to create employment and as an instrument to deliver hidden subsidies to parts of the economy.

12.1.3   Phase III: late-20th century – unbundling, competition, regulation and privatisation

The economic costs of public ownership and monopolistic market structures became more and more apparent. In the 1970s, the United States began to experiment with power sector reform. By the 1980s, policy makers in Europe, the Americas and elsewhere realised that electricity, natural gas and telecommunications were no longer natural monopolies. Thanks to advances in technology, economic theory, and increasingly sophisticated regulatory instruments, it became feasible to introduce competition with the same effect as in other industries. Substantial improvements in operational and investment efficiency, the reduction of costs to end-users, an improvement of services, and a higher rate of innovation thus became possible. During the 1990s, electricity and natural-gas sectors have been transformed through the overhaul of regulatory frameworks, the introduction of competition, and increasing private participation. These policy reforms have been implemented in developed and developing countries alike.

12.1.4   Phase IV: recent developments – industry convergence and globalisation

The fourth phase, which is now overlapping with the third, is characterised by convergence in the electricity, natural gas, and more generally the utility sector. 'Multi-utilities' are being formed to offer comprehensive service packages to clients and reap the associated economies of scale. As liberalisation and privatisation are taking hold, the industry is rapidly globalising through international mergers and acquisitions, cross-border trade, and the creation of regional power pools. Another facet of the fourth phase is the emergence of a new 'service' sector in the power industry, quite distinct from physical distribution, classified now as the 'wires' business, involving electricity markets, and trade.

12.2   Reorganising the ESI

The ESI worldwide is being restructured, liberalised, privatised, corporatised and deregulated [2]. The meaning and differences of these terms [3] are explained here.

Restructuring. A broad term, referring to attempts to reorganise the roles of market players and/or redefine the rules of the game, but not necessarily deregulate the market.

California, for example, restructured its market, deregulated its wholesale market by lifting nearly all restrictions, but kept its retail market fully regulated. Many problems ensued.

Liberalisation. This is synonymous with restructuring. It refers to attempts to introduce competition in some or all segments of the market, and remove barriers to trade. The European Union, for example, refers to their efforts under this umbrella term.

Note that 'liberalisation' is essentially a misnomer. No electricity market has been (or, in fact, can be) fully deregulated. Experience suggests that even well functioning competitive markets need a regulator, or as a minimum, a market monitoring and anti-cartel authority. Germany is the only major country attempting to do without a regulator. Even in this case, there is an anti-cartel office, monitoring the behaviour of the market participants.

Privatisation. Generally refers to selling government-owned assets to the private sector, as was done in Victoria, Australia, and in England and Wales. It must be noted that one can liberalise the market without necessarily privatising the industry, as has successfully been done in Norway. The experience in New South Wales, in Australia, has been a mixed success.

Corporatisation. Generally refers to attempts to make state-owned enterprises (SOEs) look, act and behave as if they were for-profit, private entities. In this case, the SOE is made into a corporation with the government treasury as the single shareholder. For example, former SOEs in New South Wales, Australia, have been corporatised. They vigorously compete with one another, while all belong to the same, single shareholder, namely the Government of NSW. The Islamic Republic of Iran has been considering such a move for generators.

Liberalising (restructuring) splits off two lines of the power business that until now have been controlled by the monopoly utilities – the generation of electricity, and the billing and metering of it – and allows new players to compete in providing those services. Further, it allows these competitors to set their own prices, rather than negotiating with state regulators on a fixed rate.

Only a handful of companies are competing to provide billing and metering services. But many are looking to be the biggest and the best at owning and operating power plants. Those players are usually unregulated subsidiaries of holding companies for traditional utilities, or independent companies that own only power plants.

Eventually, these companies will sell most, if not all, of the electricity into the wholesale market, where utilities and power marketers that distribute to retailers will purchase the power. The wholesale market is growing, and is considered crucial to liberalisation because buyers will pay prices based on supply and demand. That means, liberalisation proponents argue, that power-plant owners will attempt to operate the most efficient plants possible, and then sell electricity at the cheapest rate to marketers. They also argue that, currently, utilities have little incentive to operate efficiently because they get a guaranteed rate of return from regulators based on how much the plant costs to build and maintain. This is true but only to a certain extent.

As we will see there is a need for an alert regulator to avoid market volatility and ensure fair play of the markets. Disaggregation of the value chain and benefiting of new technologies, particularly in information, helped creating electricity trading and markets. Liberalisation and technology advances are driving industry transformation and a new vision of the digital utility, which is able to gain business advantage through:

- focusing on a few segments of the energy production and delivery value chain such as generation, transmission, customer care, and billing or new areas like energy trading and risk management; and
- achieving efficiencies through economies of scale gained from concentration on building best-in-class, core, mission-critical, business systems and partnering, outsourcing, or procuring the other less critical or more contextual business services required for business operations.

While liberalisation may be viewed as a driving force, the power of software and the emerging capabilities for increased collaboration and integration through open Internet-based standards are enabling and simplifying this transformation. A truly excellent software platform and applications environment provides the ability to:

- visualise, by delivering the right information in real-time where needed within a personalised portal view on any device including mobile wireless devices;
- optimise, by taking this information and combining it with other information seamlessly to get a complete picture, finding individuals with expertise that one can collaborate with, and using the information to do analysis; and
- automate, by executing processes and transacting business, which may require orchestration among many systems and processes and interoperability with many different heterogeneous systems.

12.2.1   The opportunities presented by liberalisation

Figure 12.1 outlines five possible (but not exhaustive) models for the future organisation of the electricity industry [2]. The five models range from the most conservative, which enables all players in the industry to intervene (model 1), to the solution that rules out all intermediate stages (model 5). A distinction is made between network distribution (physical transportation of electricity) and marketing distribution (supply). Third-party access (TPA) enables all players in the industry to be potential suppliers, whether they are producers, transporters, network distributors or marketers (physical traders). Marketers have a unique status, as they are the only participants to focus on supply, since they buy electricity in bulk and sell it to eligible clients.

In model 1, eligible buyers are supplied by marketers, who themselves are supplied by (network) distributors. The distributor buys the electricity from the transporter, who is the only player to have direct contact with the producer.

In model 5, eligible buyers skip all the intermediate stages between themselves and the producer, and negotiate their supply directly with the producer. This can also happen in reverse order, i.e. the producer may directly approach an eligible

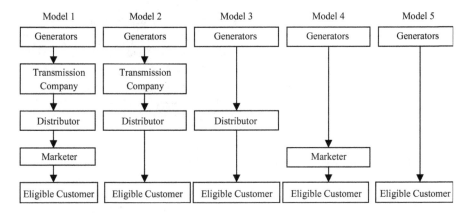

Figure 12.1   Possible models for the organisation of the electricity industry

customer. Under this model, the transporters and distributors only handle the physical transportation of electricity, as stipulated in the directive.

There are several intermediate stages between these two extremes – three of which are models 2, 3, and 4. Model 2 represents the situation that dominated the European electricity industry for a long time, with two slight differences: first that the producer was often also the transporter, and even the distributor; and, second, that some major consumers were supplied directly from the transporter's network.

With liberalisation, the three main players (producers, transmission and distributors, marketers) are likely to have different fortunes.

(1) *Enhanced opportunities for producers.* Producers have the advantage of controlling the resource in question. Liberalisation of the market will enable them to sell directly to consumers and to maximise their profits, as they will only pay the transmission company and the distributor the cost of using the networks.

Alongside these opportunities in terms of direct contact with the customer, electricity producers are benefiting from increasing synergies between natural gas and electricity at the production level.

(2) *New challenge and reduced role for transmission companies and distributors.* At first sight, transporters and distributors are the main losers from liberalisation. As they are forced to open their networks to competitors, they will lose control of a key link in the chain, which afforded them control over supply. The challenge facing these players is to combine their traditional expertise in network management (and production in certain cases) with expertise in supply. This is the condition that will enable transporters and distributors to retain their position as key players in the market, capable of handling their own electricity production as well as that bought from other producers on increasingly competitive markets. In this respect, the opening of the markets constitutes a sizable opportunity. These opportunities are local as well as global.

(3) *Traders as the link between producers and consumers.* The main advantage that these traders have regarding final consumers is their understanding of the market (i.e. their knowledge of the cheapest sources of supply) and their capacity to buy large volumes of electricity, which provides them with access to competitive supply tariffs. Similarly, producers, who do not have the expertise needed to ensure sufficient penetration of consumer markets, will prefer to work with external traders who are able to offer such market penetration.

12.3   Barriers in liberalised markets

Monopolies exist because other firms find it unprofitable or impossible to enter the market. Correspondingly, monopolies exist because of barriers [4]. However, in order for perfectly competitive prices to develop, fundamental assumptions of competitive markets must be met. One of these assumptions – the ease in which firms are able to enter markets – plays an important role in the development of competitive markets. Market entry assures (1) that long-run profits are eliminated by the new entrants as prices are driven to be equal to marginal cost, and (2) that firms will produce at the low points of their long-run average cost curves. Even in oligopolistic markets, long-run profits and prices exceeding marginal cost can be eliminated if entry is costless.

The recent California experience has highlighted the full extent of barriers facing new generation, and the cost to society when entry is constrained. However, until the barriers to entry are relaxed, prices will not be set at marginal cost. Because entrepreneurial merchant generation is unable to quickly enter the market to capture excess rents, existing generation is able to charge prices exceeding marginal cost.

There are several reasons why entry is constrained, including site development and permitting delays, turbine availability and construction lead-time. Both advanced and conventional combined-cycle technologies require three years' construction lead-time, while coal and nuclear plants require four years and more. Once the facility is built, transmission rights and fuel availability constraints can limit market participation. Finally, scheduled maintenance and physically operating constraints can limit real-time market participation. It is apparent that physical generation by itself will not provide real-time market entry and exit required to assure marginal cost pricing.

12.4   Electricity and the new digital economy

The 20th century was characterised by the industrial analogue economy. The economy of the 21st century is the networked digital economy [5], which only runs on electricity characterised by a real-time flow of information. Electricity-based innovation lies at the heart of economic growth.

Digitisation of the global economy has proceeded in three phases. First came computers, which revolutionised information processing and fundamentally transformed the way most businesses operate. Next, as the cost of microprocessors plunged, individual silicon chips began appearing in many applications – from industrial process equipment and medical instrumentation to office machines and home appliances. Now, phase three involves linking these computers and microprocessors together into networks. There are, at the beginning of the 21st century, more than a million Web sites available on the Internet, potentially available to more that 200 million computers around the world. Eventually, many stand-alone microprocessors will also be linked to networks, supplying critical information on equipment operations and facilitating even more profound changes in daily life.

This proliferation raises two challenges: quantity and quality. Quantity has to be met not only in building new power stations and generating plants but also in enlarging and strengthening the transmission network, building new interconnections, as well as enhancing the distribution system. Quality was dealt with in Chapter 11.

In 2001 the information technology (IT) itself accounted for an estimated 13 per cent of the electrical energy consumed in the US, a proportion that may grow to as much as 50 per cent by 2020, as shown in Figure 12.2.

The new digital-quality power needs are as explained above in terms of quantity as well as quality. Such needs can be fostered by:

- leveraging the advantages of distributed resources,
- defining and facilitating value-added electricity services,
- providing new d.c. electricity supply technologies,
- developing and employing advanced power conditioning, power quality devices, and power electronics, and
- establishing new service quality standards for electricity and related products.

Most of these needs have already been discussed in the book.

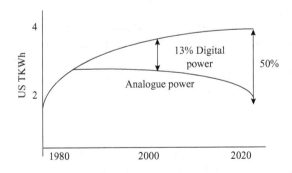

Figure 12.2   *Information technology accounts for an estimated 13 per cent of the electrical energy consumed in the United States, a proportion that is growing rapidly*

12.5   The utility of the future

Liberalisation provides for third-party access to the electricity grid and the establishment of independent grid operations [5–7]. In doing so, it will also provide for increased retail competition, starting with the largest electricity customers. Another feature is the unbundling of the vertically integrated value chain, creating different roles for current utility industry entities and allowing new players to enter the market.

The utility companies of the past ran the entire process, from generation to retail settlement. The new unbundled value chain breaks out the functions into four basic entity types; different factors define success within each of these four distinct functions.

- Generators will need to optimise generation outputs according to market demand, comply with environmental rules, prepare for new investments, and decrease maintenance and operation costs.
- Energy network owners and operators will concentrate on new ways to secure physical delivery, decrease maintenance and operation costs, plan new investments, and ensure compliance with regulatory authorities.
- Energy traders, brokers, and exchanges should focus efforts on creating trading vehicles and market liquidity, managing risk exposure, lowering transaction costs and developing new wholesale offers that make the best use of connectivity technologies.
- Energy service providers and retailers will establish ways to reduce energy-sourcing costs, manage client portfolios and risk profiles, bundle and market new services, and launch innovative offerings to large clients.

Liberalisation forces a disaggregated business model for each separate element of the value chain, even if the specific models differ from country to country and region to region. This disaggregation promotes the inclusion of new participants with a new combination of roles into the market, participants (such as asset managers, wholesalers, traders, and transmission system operators) who did not exist a few years ago in the utilities landscape.

Many of the structural changes that must take place during the transition will be driven by new technologies, such as electronic exchanges and online trading. Today's focus is on external communication needs to interconnect all of the participants, including suppliers, customers, and partners, in the value chain.

The large utility of the future must be able to thrive in a competitive, global market for energy [7]:

- it is multinational,
- it can carry its expertise over into emerging parallel businesses,
- it has the flexibility and willingness to unbundle,
- it adapts readily to new structures and concepts,
- it goes beyond its traditional geographic borders to grow, and
- it is an expert manager of risk.

The utility has to learn to operate in different cultures, regulatory environments and markets. The big challenge is in the huge transitions that occur when the industry is transformed by privatisation, consolidation, liberalisation, and the introduction of real free-market competition, not to mention the convergence of energy, capital, and reinsurance.

The utility of the future will have the flexibility and willingness to unbundle whenever doing so can add value and competitive advantage as well as improve service to the customer. Risk management will be more important than ever in tomorrow's more deregulated energy markets because it protects businesses, municipalities and other entities affected by change in the energy market, the weather, or other factors. The utility of the future will manage commodity price risk with options, swaps and other derivatives. Increasingly, it will be bundling energy, capital and actuarial risk products to manage a broader array of client risks. The benefit is that it will be able to manage risk for itself and its clients, commercialising or productising risk management services. The utility of the future will be an expert manager of risk and help customers determine and achieve an acceptable risk level.

12.6 The 'virtual utility' of the future

Gradually more distributed generation is going to evolve, and most of it will be small units connected to the grid. Such small and decentralised electric energy sources can be operated over the Internet as one large generating plant, a 'virtual utility' [8]. Such development is being enhanced by environmental considerations that are encouraging small renewable energy sources, the fastest growing of which is wind power and small hydro, microturbines and in the future, fuel cells. Such systems may be interconnected by d.c. transmission systems through low-cost cable network.

This is accompanied by a trend towards neater and aesthetically more appealing arrangements of the network; underground cables are growing while overhead lines are shrinking in size. Cable technologies are allowing for economical and easy-to-install systems. Substations are becoming more compact and easier to conceal. Major advances and cost reduction are also taking place in HVDC light technologies.

These multi-sources will need extended and sophisticated control of all the components and the power flow. The decentralised electric power generation and related revenue flows will most likely be handled by the Internet for direct control, and also for maintenance of the grid as well as for financial and administrative transactions.

12.7 DSM programmes in deregulated markets

In regulated markets, the cost and responsibility of demand side management (DSM) programmes were built into the rate-base or funded through green energy surcharges. In deregulated markets, where DSM programmes or renewable energy investment must be recoverable through market-based pricing, these programmes have been considered uneconomic and thus neglected [4].

Theoretically, real-time load management is analogous to physical ancillary generation markets. Rather than dispatching and curtailing generation, real-time load management curtails and dispatches load. However, owing to the high cost of monitoring and telemetry equipment and current limitations in market design, practical real-time load management is only available to large industrial consumers.

However, residential consumers can also participate in load-curtailment markets. Residential customers can be encouraged to shift demand from peak to off-peak hours via a multi-tier tariff. For example, a simple two-tier system that prices peak power consumption differently from off-peak would provide incentives to shift non-essential activity to off-peak hours. Although limited, the opportunities for residential consumers provide a significant potential source of peak-load reduction. However, the current system of load profiling is fundamentally inconsistent with real-time load measurement and pricing.

Demand responsiveness markets will be most effective when shedding peak load. Experience has shown that a small demand reduction could effectively bring wholesale prices way down. In many service territories, peak demand for the system, which may represent only 100 h or so per year, creates the need for 10–25 per cent greater system capacity. For peak-load shedding markets to develop, peak-load price signals must be passed to end-use customers. As price signals become apparent, more end-users will find the flexibility and desire to sell back megawatts into the grid.

The most successful programmes avoid much of the downside price risk through voluntary participation. Instead of threatening users with the possibility of extreme energy costs, voluntary programmes pass the price signals to the consumer and, therefore, the incentive to curtail. However, if the consumer chooses not to respond and continues current consumption, they pay the conventional stable rate for electricity. Under voluntary load curtailment, the energy user pays a standard rate that is designed to average out the highs and lows but, during a price spike event, the user can 'sell back' the curtailed energy to the electricity supplier.

Ideally, the electricity supplies would be indifferent to either paying the generating company the spot market price for wholesale energy or paying the large load reducing negawatt participant for load curtailment. Under this scenario, the end-use customer receives the full benefit of equivalent spot market prices for participation in the negawatt market. The benefit to the supplier is less apparent. If the load curtailment generates enough savings, the market would face a less-expensive marginal unit setting market price. In this case the supplier would receive a higher return on power sold to fixed tariff customers.

Load responsive negawatt markets can provide system capacity through either reducing consumption or switching to backup-generation. For the purpose of calculating the cost to shed system load, the two options are equivalent. Both switching to backup-generation and shedding load represent opportunity cost. However, the advantage of focusing on the cost of backup-generation is that it effectively sets an upward bound on cost. The annualised cost of backup-generation effectively caps the power market annualised price. At the point where system cost exceeds the cost of new generation, negawatt market participants would be better off installing new backup-generation than purchasing from the power market. Negawatt markets would compete directly with the generating company, creating a demand response cap to market price and volatility.

Although negawatt market participation can be either through reducing consumption or switching to backup-generation, for the purpose of market pricing, we consider all participation as if through backup-generation.

Participants in an energy reduction programme receive a corresponding premium payment and an energy credit for curtailed energy. The premium payment is based on the 'strike price', the option load contracted and the operational plan selected. A 'call option' in this case gives the supplier the right to purchase energy from the end-use customer at the agreed upon strike price. The call option is exercised when the supplier marginal cost of electric energy, including all variable cost associated with delivering the energy, is projected to be equal to or greater than the strike price.

Theoretically, real-time load management is analogous to physical ancillary generation markets. Rather than dispatching and curtailing generation, real-time load management curtails and dispatches load. Responsive load negawatt markets can be developed to create real-time entry and exit fundamental to competitive priced electric power markets. Negawatt markets would compete directly with generating companies, creating a demand response cap to market price and volatility. Generators would compete with backup-generation, the cost of which sets the market cap.

Using a market-based method of pricing real-time load curtailment, based on real-option valuation of participant opportunity costs, price incentives exist for negawatt market development. The strike price is given by a contractually agreed upon threshold price between the energy provider and energy consumer. From price volatility determined from historic price data or implied from forward markets, a premium value is calculated for the right to curtail future load. Option premiums, profit sharing and limit orders can provide financial incentives for functioning demand responsiveness markets.

12.8 The need for a regulator

It is thought that liberalisation will eliminate the need for a regulator [9]. It is also thought that powerful price signals – emanating from fluctuations in supply and demand – are generally adequate to regulate competitive markets. There are cyclical shortages, which result in higher prices, which encourage increased supplies and, which bring prices back to normal levels. So self-correcting competition, obviates the need for a regulator to monitor prices and signal when additional supplies are needed and so on. Such arguments do not seem to apply adequately to electricity markets.

An alert and all-powerful regulator appears to be needed particularly during the critical early years of market reform because the market may not function properly from the beginning. But even after the initial difficulties in the system are worked out, there appears to be a continuing role for the regulator to enforce the rules, and to ensure competitive behaviour among market participants.

One of the key questions for policy makers and regulators is whether there are adequate incentives for investment in generation, transmission, and distribution networks in competitive electricity markets. This is an issue that a regulator has to monitor, providing encouragement and incentives for the market to invest to cover for a gradual increase in demand. However, encouraging adequate private investment in transmission is not easy. Costs are reasonably well known, but benefits are difficult to measure and harder to collect because they may accrue to third parties. An alert regulator will have to monitor the network and ensure that there are rewards and incentives to invest in transmission strengthening.

12.9   References

1   'The Development of Electricity Markets in the Euro – Mediterranean Area', Daniel Muller – Jentsch, The World Bank Technical Paper 491. The World Bank, Washington, D.C., 2001

2   NYOUKE, E.: 'Deregulation and the Future of the Electricity Supply Industry'. *World Power,* 2000, pp. 34–38

3   SIOSHANSI, F.: 'Sobering Realities of Liberalizing Electricity Markets', *IAEE Newsletter*, Third Quarter 2002, pp. 24–32

4   SKINNER, K.: 'Market Design and Pricing Incentives for the Development of Regulated Real-Time Load Responsiveness Markets', *IAEA Newsletter*, First Quarter 2002

5   LEWINER, C.: 'Utility of Future – Business and Technology Trends', *IEEE Power Engineering Review*, Dec 2001, pp. 7–9

6   GELLINGS, C., and SAMOYTI, M.: 'Electricity and the New National Economy, Electric Infrastructure to Power a Digital Society', *IEEE Power Engineering Review*, January 2002

7   GREEN, R.: 'Twenty – First Century Utilities', *IEEE Power Engineering Review*, Dec 2001, pp. 4–6

8   BAYEGAN, M.: 'A Vision of the Future Grid', *IEEE Power Engineering Review*, Dec 21, 2001, pp. 10–11

9   IEA.: 'Investment and Security of Supply in Electricity Markets', 2002

Chapter 13

Project analysis: evaluation of risk and uncertainty

13.1 Introduction

From a practical point of view there is no project without risks. Risk taking is normal to entrepreneurs, to lending and funding agencies and also in government development plans. Such risks and their extent are reflected in choosing the discount rate of the project (see Chapter 4), where investors expect higher returns to compensate them for risk taking. Because of the regulatory nature of the industry, the limited number of players and the unique nature of electricity and its continuous rise in demand, the average project in the electricity supply industry is less risky than the average investment in the stock exchange [1,2]. However, each project has its own risk; projects that involve new technologies (renewable and clean-coal technologies), or projects with lengthy lead times (nuclear and hydro-power) involve a lot of investment and have more than the average level of risk. Some of the risks are related to engineering and technology; however, market risks equally exist. For example, the demand may not turn out to be as estimated, the tariff is lower than expected, project execution may take more time and involve more cost than planned. Future fuel prices are one of the most risky aspects in evaluating investments in the electricity supply industry projects. Variation in fuel prices may surpass expectations, or supplies may turn out to be insecure and more expensive alternatives have to be sought. For projects with lengthy lives, like coal-firing and nuclear power stations, the problems of obsolescence (due to technological change) and environmental legislation exist; these may cause such projects not to survive their full life or end up with heavy and expensive modifications. Cost overruns, which are caused by project delays, or inaccuracies in estimation do not only significantly change project costs but also substantially reduce net benefits.

13.2 Project risks

Project evaluation involves assumptions about inputs with varying degrees of uncertainty; in some cases these individual uncertainties can combine to produce a total uncertainty of critical proportions [3]. The risks can come from three sources: uncertainty in project planning and specifications, uncertainty in the design coefficients (engineering and technology, economic and income prediction, etc.) and uncertainty in exogenous project inputs (mainly fuel availability and prices). Sensitivity and risk analysis affect the choice of the least-cost solution as well as the IRR and correspondingly the decision to proceed with the project altogether.

Most of the financial and project evaluation of Chapters 5 and 7 was on a *deterministic* basis, i.e. all inputs of costs and benefits and corresponding cash flow forecasts were assumed to be accurate. Risks were only incorporated in the discount rate and in allowing for contingencies in project cost, and when net present values were calculated. As already explained, in the real world things never work out that nicely. Therefore it is prudent to account for such uncertainties by carrying out some project risk analysis. Such analysis involves the following.

- Evaluation and judgement about the behaviour of certain uncontrollable inputs of the project. Calculating, instead of a single return, a whole set of possible returns for the project based on the behaviour of the uncontrollable inputs.
- Criteria for choosing the least-cost solution among the different alternatives based on the likely set of returns for each. A less-risky project alternative with modest returns may be preferred to a more-risky alternative with probably higher returns.

Apart from giving a more realistic and accurate assessment of the likely outcome of the project, a risk and uncertainty analysis has many other advantages. It enables more accurate analysis and evaluation of project inputs and outcomes by many experts and with more factual evidence. This will also lead to an insight into restructuring or redesigning the project in light of this evaluation and analysis. It also enables focusing on those uncertain inputs that affect the likely outcome so as to mitigate or reduce their uncertainty in a number of ways. Project analysis is greatly assisted by modern computers and statistical analysis packages, which allow undertaking of such evaluation in a short time and at a limited cost. However, it needs a lot of insight and accurate understanding of the project structure and details.

Uncertainties faced by power utilities are both *internal* factors, which can be controlled even to a certain extent by the utility, and *external* factors, which are outside the control of the utility (see Table 13.1). Internal factors can be managed by better handling of the projects' design, execution and operation; outside factors have to be incorporated into the planning process in order to reduce their uncertainty. The experience of planners and designers greatly assists in defining uncertainties and their handling. Post-evaluation of projects to compare the planning assumptions (costs, execution period, demand prediction, etc.) with actual events, builds experience that helps planners to reduce risk and uncertainty in future project assumptions and plans. For instance most power projects financed by the World Bank proved to cost more and take lengthier times to implement than planned, and detailed sensitivity analysis does not adequately capture or account for such differences [4].

Table 13.1   Major uncertainties faced by electricity utilities [4]

External uncertainties
- national and regional economic growth and its corresponding effect on future electricity demand,
- structure of energy demand and rate of substitution of other fuels by electricity,
- unpredictable future fluctuations in local and global fuel prices,
- insecurity in imported fuel sources,
- future environmental regulations and legislation,
- technological innovations and development of more efficient and clean plant,
- future inflation and cost of financing.

Internal uncertainties
- project cost overruns,
- project execution schedules and delays,
- reliability of the system and availability of generating plant,
- system losses,
- operation and maintenance costs.

There are many ways to handle and deal with uncertainty and risk. These can be grouped under the following headings:

(i)   delay and defer decisions until the uncertainty is reduced,
(ii) plan and allow for possible short-term contingencies,
(iii) sell risk to others through turnkey projects, long-term fuel contracts, etc.,
(iv) flexible strategies that allow for relatively inexpensive changes like designing the system to allow for possible future fuel conversion, executing the network to allow for operating at higher voltage, etc.

Therefore recent trends in generation planning try to avoid investing in large units with long lead times and concentrate instead on smaller sets with short construction periods and limited investment. This will greatly reduce capital risks of utilities, as already explained in Chapter 1.

For simple and small projects, uncertainties can be adequately evaluated by a simple sensitivity analysis, through measuring the response of the IRR or net present value to predictable variations in the project inputs. Alternatively it may need a full risk analysis utilising a *probabilistic* approach to assess the combined net effect of changes in all variables or the likelihood of various changes occurring together. Such a probabilistic approach is particularly useful, indeed necessary, in the case of large capital intensive and risky projects.

Generally speaking there are several procedures for project risk analysis, mainly: sensitivity analysis, decision analysis, break-even analysis and the Monte Carlo simulation [5].

13.3   Sensitivity analysis

Sensitivity analysis involves calculating cash flows under the best estimate of input variables, and then calculating the consequences of limited changes in the value of these inputs. It assists the evaluator to identify the variables that significantly affect the outcome and correspondingly needs more information and investigation. Sensitivity analysis is carried out during practically every financial and economic evaluation of projects; it is simple and informative. It is a review of the impact that changes in selected project inputs, costs or benefits, or a combination of these, can have on the project's net present value or IRR. In this case one or more variables are changed independently or collectively, within reasonable limits (say, 10–20 per cent) to see the likely effect of the change on the net present value of the project and its likely IRR. Alternatively the aim may be to calculate the change in one variable, like the selling price per unit or project cost that will reduce the net present value of project benefits to zero. This will indicate the selling price per unit or project cost below or above which, respectively, it is not worthwhile pursuing the project. Another alternative may be *break-even analysis*, to work out values of inputs and outputs that will reduce the project benefits to below the *cut-off rate*, which is a rate established as a 'threshold' below which projects should not be accepted. The cut-off rate depends to a large extent on the riskiness of the project. A project with a high volatility in some inputs will therefore need a much higher acceptable cut-off rate than a project with almost sure estimates.

One of the most important exercises in sensitivity analysis is to review the impact of the discount rate on the project's net present value and its profitability. In this case, and if there is no single firm discount rate, more than one discount rate is tried and the outcome with each discount rate is described. The UNIPEDE/EURELECTIC study into the projected cost of generating electricity considers two discount rates, 5 per cent and 10 per cent, each applied with different outcomes [6].

Sensitivity analysis is therefore an essential and easy means of evaluating the vulnerability of the project to likely future deviation from best-input estimates. It can also greatly help in assessing the extent of risk in the project, and the particular inputs that significantly affect the project outcome. Once these are identified, then a more careful study should be undertaken of these particular items to enable better estimates and a firmer calculation of the net present value and the project's IRR. For the electrical power industry the most important items affecting a project's financial performance are the electrical tariff and fuel prices. With the increasing availability of electronic calculating facilities it has become much easier to undertake many sensitivity analysis scenarios and to analyse the effect of various parameters on the project's financial and economic feasibility.

One of the major weaknesses of sensitivity analysis is that the changes are, most of the time, *ad hoc* (10 per cent change in price of fuel or 10 per cent change in demand, etc.), without regard to the expectancy and probability of these happening. Such *ad hoc* assumptions do not assist decision makers to examine the likelihood of the event. Fuel price changes, in the future, may be much more likely to occur than other operational costs, therefore dealing with these two on equal terms can lead to wrong impressions and conclusions. So also can the effect of a change of one input on other inputs; a significant change in fuel prices will not only affect the NPV and IRR, but will also affect demand, prices and merit order and can cause significant implications, which are not captured by sensitivity analysis. What is important is not only the prospect of change of a fundamental input assumption but also the probability of this happening and its extent, and also the interrelationship between variation in one variable and other inputs. Such prospects can only be adequately evaluated by proper risk analysis.

To demonstrate this, consider a 100 MW CCGT set, firing LNG, that consumes 6000 BTU per kWh generated, at a cost of £2.60 per million BTU. The set will cost £50 million at commissioning and is expected to act as a base-load generating unit at full load for 7000 hours annually. There is a fixed annual cost of £1 million.

With a 10 per cent discount rate, over 20 years, the equivalent annual cost of investment is equal to

$$\frac{\text{investment}}{\text{20-year annuity factor}} = \frac{\text{£50 million}}{8.514} = \text{£5.87 million.}$$

The fixed annual cost of capital and operation will equal

$$5.87 \text{ million} + 1 \text{ million} = \text{£6.87 million,}$$

with an annual generation of

$$100 \text{ MW} \times 7000 \text{ h} = 700 \text{ GWh.}$$

The fixed annual cost per kWh will be

$$\text{£6.87 million} \div 700 \text{ GWh} = 0.98\text{p.}$$

Fuel cost per kWh is
$$(\text{£2.6} \times 6000)/10^6 = 1.56\text{p.}$$

Total cost of generation is

$$0.98 + 1.56 = 2.54\text{p kWh}^{-1}.$$

Consider the case of a sensitivity analysis of a possible increase in fuel cost by 20 per cent, then the fuel cost will become 1.872p kWh^{-1} and total cost of generation 2.85p kWh^{-1}.

With such a fuel price and operational cost, the set most likely will not remain a base-load set and correspondingly the energy output of the set and its cost per kWh will be higher (depending on the position of the set in the merit order). Sensitivity analysis fails to capture such correlation between fuel cost, energy contribution of the set and correspondingly system cost. It is only simulation that can give the generation cost, capture the system effect and evaluate the true economics of investing in such a plant compared with other alternatives firing other fuels.

13.4   Break-even point analysis

In all industrial production projects a financial break-even point analysis is essential in order to assess the relationship between production volume, production cost and profits. The break-even point is the level of product sales at which financial revenues equal total costs of production; at higher volume of sales financial profits are generated. The detailed break-even analysis will vary depending on what type of profits (gross, net, before or after tax, etc.) and what type of costs (cost of money, cost of certain input items) it is required to test. It is usual to carry out such testing through sensitivity analysis in order to demonstrate how variation in different components of production cost or demand affect this break-even point.

Break-even point analysis is essential in industrial projects. It can be carried out rather easily by equating fixed costs and variable costs to income from sales at a certain sales level so that

fixed (capital and fixed operation) costs

+ (sales volume × variable production cost per unit)

= sales price × sales volume.

One important point to remember in break-even analysis is that it has to be carried out through proper financial evaluation, like that in Chapter 5, rather than through financial accounting statements, which is a common error that most firms fall into.

For example, consider a firm that will invest £10 million to produce a product that has a market price of £2 per unit and a £0.5 variable cost per unit. The firm will face £0.25 million fixed expenses annually. The business is going to remain for 10 years and the company employs a straight-line depreciation method, where sales equal production. It is required to find out the break-even point.

The annual fixed costs of the company are

depreciation (£10 million per 10 years)	£1.00 million
other fixed costs	£0.25 million
total	£1.25 million.

The break-even sales are now calculated according to the earlier equation:

£1.25 million + (£0.5 × volume of sales) = £2 × volume of sales

£1.25 million = £1.5 × volume of sales

break-even volume of sales = 833 thousand units.

Therefore this firm will go into business only if it is guaranteed that its sales will exceed 833 thousand units, on the average, annually.

This accounting approach basis is quite misleading since it ignores the opportunity cost of capital. As mentioned in Chapter 6, depreciation is accounting cost that should not go into financial evaluation. The real cost of money is its opportunity cost; assuming that this is 10 per cent, then the equivalent annual cost of the investment is:

$$\frac{\text{investment}}{\text{10-year annuity factor}} = \frac{\text{£10 million}}{6.145} = \text{£1.627 million}$$

annual fixed cost $= \text{£1.627 million} + \text{£0.25 million} = \text{£1.877 million}$

£1.877 million $+$ (£0.5 $\times$ volume of sale) $= \text{£2} \times$ volume of sales

break-even sales $= 1251$ thousand units.

(This new break-even volume of sales is more than 50 per cent higher than the earlier figure, which displays the importance of financial evaluation and having proper assumptions. Depreciation does not reflect the true cost of money and it is quite misleading in evaluation in that it leads to a lower break-even point.)

To assess the effect of production cost per unit, a sensitivity analysis is carried out by increasing this by 20 per cent, to £0.6 per unit, and measuring the effect on the break-even sales volume. This will mean increasing the break-even sales volume into 1341 thousand units, i.e. 7.2 per cent more.

Sensitivity analysis, as has been proven by experience, is not adequate for large capital-intensive projects. However, it remains a simple and widely used tool for assessing the risk in project valuation and in drawing the attention of planners to particular project inputs that warrant special attention, so that these can be investigated better, or so that certain arrangements are made beforehand, to reduce their risk. Sensitive analysis through varying the discount rate (the opportunity cost of capital) can have great impact on the outcome particularly when comparing a capital-intensive project with limited-future long-term operational cost (nuclear) with less capital-intensive alternatives, but higher cost of operation (CCGT and coal plant). It is much wiser to undertake an extensive exercise to find the right discount rate commensurate with the risk, as explained in Chapter 4, rather than to undertake a discount-rate sensitivity analysis, which may lead to confusing results.

13.5 Decision analysis

A more comprehensive approach to uncertainty than sensitivity analysis is decision analysis [7]. In its simplest form it involves selecting scenarios: the base scenario, an optimistic scenario and a pessimistic scenario. The base scenario will demonstrate the most likely inputs and outputs from the point of view of the project evaluator. The optimistic scenario will incorporate future parameters that are more favourable to the project success than they appear at the time of evaluation, while the pessimistic scenario will have an opposite view expecting that future events may be less favourable than they appear today. A decision analysis can be undertaken where each of these scenarios can be assigned an explicit probability and be represented as decision tree. A question about the discount rate to be applied arises; a pessimistic scenario will have much less risk than an optimistic scenario, so will the same discount rate be applied in both cases? This is a valid argument that will be dealt with later.

Detailed decision analysis involves the construction of a set of mutually exclusive scenarios of an event, the sum of probabilities of these scenarios adding up to unity. Only one of these mutually exclusive scenarios must take place. Decision analysis is a way of dealing with prospects for future variation of key inputs. Each input can have exclusive scenarios with these scenarios covering all possible future states (N), each state with a probability (P), such that the sum of all these probabilities equals unity ($\sum_N P_n = 1$). If the possible values of each input quantity are small, then uncertainty can be expressed as a discrete probability distribution of the output.

The probability of the input quantities can, in most cases, be better presented as continuous, rather than discrete, distribution. It is possible, as an approximation, to express a continuous probability function by a discrete distribution and use the probability tree approach. But with many input variables, proliferation of the tree will occur, and this makes the analysis rather cumbersome.

Setting up mutually exclusive scenarios with a probability distribution assigned to each is a rather difficult task. These are usually based mainly on the judgement of experts and other project designers. It involves a lot of estimation and future guessing. Decision analysis is an improvement on sensitivity analysis; however, its weakness lies in the selection of discrete assumptions and probabilities, which are arbitrary in many cases. Also the method does not capture the covariation that may exist among different variables.

Decision analysis, therefore, is a step forward from sensitivity analysis, yet it has its shortcomings, which have already been explained. However, if the likely input probabilities are few and discrete and can be presented in a simple decision tree, then this can assist in better risk evaluation of the project. Such decision trees are very helpful in sequential decisions. The sophisticated approach to risk assessment is to characterise the uncertainty in the project inputs by assigning a probability distribution to each important input and to simulate the project output – IRR or net present value – by a probability distribution as described in Section 13.6.

13.6   Risk analysis (the Monte Carlo simulation)

In sensitivity analysis the impact of change of one variable at a time is evaluated; decision analysis allows the evaluation of the effect of a limited number of plausible combinations of variables. Risk analysis is the tool for considering all possible combinations. As already explained, because of uncertainty about major inputs the selection of a single value for the input variables (as is the case in the deterministic evaluation) could lead to misleading outputs. Risk analysis is basically a method of dealing with uncertainty. Recent trends in the electricity supply industry, particularly the growing importance of competition and markets and independent power producers, have strengthened the need to resort to risk analysis in order to deal with uncertainty. Uncertainty is characterised by trends, events and developments that are not known in advance but are likely to happen, and whose occurrence can affect the outcome of planning decisions. The fundamental risks that are encountered in the electricity supply industry have already been mentioned. They are primarily market (demand growth, future fuel prices, cost of money, tariffs and other income prospects) and technical (breakdowns and non-availability of plant, technological change and prospects for future environmental legislation and enforcement). During project implementation there are also the risks of cost overrun, delays in implementation and teething troubles in commissioning. Also the changing structure of the electricity supply industry worldwide is creating uncertainties about regulations, the introduction of new players, the phasing out of monopolies and possible competition.

For most small projects, sensitivity analysis (or the more sophisticated decision analysis) can be enough to satisfy the investor as to the vulnerability of the rate of return to reasonable variations in the values of key inputs of the project, and hence the risk extent in the investment. However, in the case of more important investment projects, particularly in new ventures with little past experience, a proper risk analysis has to be undertaken. Risk analysis calls for sophisticated knowledge of expectancy and probability theory and the undertaking of value judgement.

In the deterministic approach, the NPV and the IRR for the project were presented as single figures. In actual fact each is a single point on a continuous curve of possible combinations of future happenings. Therefore it is much more accurate to calculate the IRR in a form of schedule (or curve), which will not only give the most likely value of the IRR but also the chance of this value happening and the chance of its being lower (or higher) than the likely value, as shown in Figure 13.1. This analysis would give the investor a much better knowledge as to the prospect of his investment achieving the expected return, or any variation of this, with the probability of it happening. This will not only sharpen the investor's knowledge of the profitability of the investment but in some cases it may also affect his choice of the least-cost solution.

The idea of risk analysis is to try to present each of the main values of inputs and outputs of the project in the form of a probability distribution curve with each event having a chance of happening. For instance the most likely cost of the project is C, but the likelihood of this happening is 50 per cent, the chance of it becoming 1.5 C is 10 per cent, but it is impossible to exceed 2 C, while the prospect of it becoming 0.95 C is 20 per cent and can never be below 0.90 C. Such information will allow the

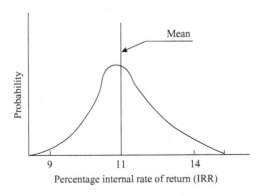

Figure 13.1   Probabilistic presentation of the rate of return (IRR)

drawing of a distribution curve similar to that in Figure 13.1. Such distribution curves can be drawn not only for the project cost but also for most of the input variables: future fuel prices, the project execution time, the cost of operation, the size of sales and the likely sale price of the output, etc.

There is no precise or standard way of drawing such curves; the curves, however, present the best-value judgement of the project planners and estimators as to the likely values concerning these main elements of costs and benefits. Such valuation depends on the estimator's value judgement, which is derived from experience and understanding of the market and knowledge derived from previous projects. By utilising the Monte Carlo method in selecting, at random, input sets of these components each with its probability of happening and calculating the IRR for each combination, and repeating this a few times, an IRR curve can be established with its peak, indicating the most likely IRR value and the chance of it happening and also the likely deviation from this return, each with its probability. Thus the extent of risk in the project becomes clear. A similar curve can be drawn out for the NPV, with each value having a probability of it happening; the same applies to similar required outputs.

The Monte Carlo simulation is a powerful tool in risk assessment. Besides identifying the important input variables, a probability distribution is assigned to each input while identifying any relationship (covariance) among inputs. Sets of input assumption are repeatedly drawn from each input distribution. This will lead to outputs characterised by probability distribution, with the possibility of calculating the mean outcomes, variances and other required parameters.

The Monte Carlo simulation involves three steps.

Modelling of the project. The complete model of the project will contain a set of equations, details and data for each of the important input variables (for instance, in the case of generation system expansion modelling – load, generation availability states, futures set sizes and set costs, maintenance scheduling, reliability criteria and cost of interruptions). The set of equations illustrates how to describe interdependence between different inputs and interdependence over time.

Specify probabilities. A probability distribution has to be drawn out for each of the inputs other than those that are known to have deterministic values (for instance fixed price fuel contracts or turnkey key projects). Such probability distribution can take one of the shapes specified in Figure 13.2 that is drawn by experts using their experience and expectations or through knowledge of past performance and data, as in the case of simulation of generation capacity states. The results are sensitive to the form of input distribution; therefore, choosing the most relevant distribution is quite important. Normal distribution is not always the appropriate distribution form unless there is supporting evidence.

Simulate cash flows and calculate output. The computer samples from each of the input variables each input value with a probability, calculating the resulting output cash flows and through them a probability distribution of NPV and IRR.

Such a simulation process is tedious and complicated; however, it has many advantages. Not only does it lead to better results with the associated probability of

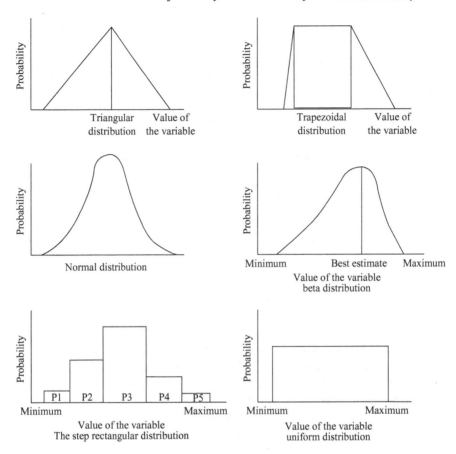

Figure 13.2   Probability distribution curves (to be used in presenting the probabilities of different input variables)

achieving these, it also enables the forecasters and planners to face up to uncertainty and interdependencies and investigate input values in greater detail. The discipline of simulation can in itself lead to a deeper understanding of the project.

Once the input distribution curves have been completed, the computer, utilising the Monte Carlo simulation, generates random values for each of the parameters, calculates the rates of return (NPV or other required output) in each run, and then repeats the proccss with new random inputs. After many runs (say 300) an output distribution is obtained with a mean and a standard deviation.

Unlike sensitivity analysis, this probabilistic risk simulation gives a complete picture of the project outputs and their chance of happening, thus quantifying the project risk. This is not necessarily the true risk that is going to occur in the future, but the risk that is the best judgement of the project evaluators and appraisers. Therefore it goes nearer to define and point out the risks of the project and its output parameters.

These, together with the mean and standard deviation, are the best definition of the project's likely output.

One of the most important values of risk analysis is that it assists in reducing the risk of the project by considering in detail inputs that increase the output risk and studying these risks with methods designed to minimise them. Sensitivity analysis can assist in pointing out the inputs that have a major impact on the project output, therefore risk analysis can concentrate on studying these. It has also to be noticed that normal contingency allowances have to be incorporated in the probability distribution.

The results of the analysis usually focus on calculating the rate of return. A few results are expected:

(1)   the mean internal rate of return and the standard deviation,
(2)   the probability of achieving a minimum return,
(3)   the shape of the distribution and the 95 per cent range confidence level.

13.7   Consideration in risk analysis

13.7.1   *Building up probability distributions*

The first and most important step in risk analysis is to assign to each input variable a suitable probability distribution. These distributions are based essentially on subjective judgement. The amount of input variables involved depends on the extent of disaggregation referred to in Section 13.7.2. The subjective judgement of the probability distribution of each input will involve a small team of evaluators and appraisers who are familiar with the inputs and their likelihood variations and also with probability simulation [8,9]. Opinions are exchanged among the team members until a probability distribution, from among classical probability distributions, is chosen as being a better fit to the case. Of particular importance are the extent of skew and the probability of the interval. The approach aims at interaction between quantitative and qualitative judgement. A quantitative judgement involves assigning tentative figures; these in turn are judged qualitatively and modified, and the process is repeated until a probability distribution curve of the input is agreed upon, by all those concerned in the process, as representing the most likely possible distribution.

As explained above the probability distribution to be utilised has to represent the best judgement of the evaluation and appraisal teams. There are many forms that a probability distribution curve can take. These are:

- step rectangular distribution
- discrete distribution
- uniform distribution
- beta distribution
- trapezoidal distribution
- triangular distribution
- normal distribution.

The form of these probability distribution curves was presented in Figure 13.2. The choice of the most appropriate distribution curve depends on the judgement of the evaluation team. There is tendency to use the normal distribution because of its neatness. Experience, however, has shown that this is probably not the best distribution curve in every case. Other forms like the step rectangular or discrete distribution are also quite useful (sometimes more useful) and easier to formulate.

Two major issues have to be considered in risk analysis: disaggregation and correlation.

13.7.2 Disaggregation

Disaggregation refers to the extent of detail in which an input has to be analysed. For instance, the cost of a power station involves civil works, turbine, generator, boiler, switchyard, fuel handling and storage. The civil works can be disaggregated into great detail of foundations, buildings, cooling arrangements, control room building, storage and workshop building, etc. Each item can be disaggregated into further details involving quantities of cement and concrete, steel reinforcements, aggregates, labour, other inputs, etc. It is important to understand where to stop disaggregating in order to construct a reasonable probability distribution for each item, and correspondingly of the project cost.

Incomplete and inaccurate results usually happen from lack of disaggregation. Through disaggregation it is possible to obtain better cost estimates, a more realistic demand prediction and a better probability distribution for the inputs. However, too much disaggregation will make the work of the evaluation team cumbersome in having to construct too many probability distribution curves. Therefore it is wise to concentrate on disaggregating those uncertain input items that can have greater impact on project viability and evaluation.

13.7.3 Correlation

Correlated variables are variables that are likely to move together in a systematic way. They are not easy to detect and are difficult to measure. They appear in practically every project and if ignored can lead to the wrong conclusions. In the electricity supply industry, demand is correlated to economic growth and the tariff; operational costs are in turn very strongly influenced by possible variation in fuel prices, this in turn is correlated with demand and economic activity.

Correlation is often hidden and difficult to detect, also it does not appear in the deterministic approach. Disaggregation helps greatly in building up and complicating the problem of correlation, and by limiting disaggregation to an adequate and reasonable level, correlation can better be dealt with. Also acquisition of more data and relating inputs can discover serious correlation that may exist and will allow accounting for this. Correlation is a much more serious problem than the choice of the probability distribution curve. It is not possible to give a general rule for the electricity supply industry, particularly the movement of fuel prices and its relationship with other inputs. Each large project has to be studied separately and its features, disaggregation and correlation adequately dealt with.

13.7.4   *Effect on the discount rate*

In the electricity supply industry different projects have different risks. Investing in a power station is riskier than investing in transmission and distribution; nuclear and new technologies power stations (including some renewables) are riskier investments than conventional power stations. All this can be incorporated in the discount rate, which can vary in accordance with the extent of risk in the project.

The second consideration is whether a closer study of the project including presenting input variables in a stochastic manner, with disaggregation and appropriate correlation, will reduce investment risk and hence lead to a reduction in the discount rate. It has to be remembered that the discount rate signifies the risk of the project. However, in spite of the care in identifying accurately the project inputs (demand, project cost and fuel prices), risk still exists. The discount rate can never be equal to the risk-free discount rate, although it will be near to it in the case of fully regulated electricity distribution utilities where there are frequent tariff adjustments.

Simulation is more realistic in predicting the future performance of the project. However, there will still be inaccuracies in representing the input variables. Also, not all variables can be accounted for and be adequately represented in a probability distribution.

Therefore it has been advocated that instead of calculating the NPV through simulation it is sufficient to calculate the IRR and represent it in a probability distribution curve similar to that of Figure 13.1 with a mean and a range containing 95 per cent of the expected values of the IRR. This will avoid the problem of assigning a previous discount rate for the project to obtain the NPV. However, if the mean IRR is computed by simulation with a standard deviation this still has to be compared with a discount rate that reflects the risk of the project as advocated in Chapter 4. Calculation of the discount rate that considers the reduced risk of the project is required.

13.8   The value of risk management in liberalised markets

Liberalisation intensifies competition; this will lead to price volatility and greater risk [10] to producers. In the past, accidents and increased costs over a short period owing to unavailability of a low-cost plant were not reflected in immediate price increases. Producers were able to delay reflecting their increased production costs in their prices. There was a time adjustment on prices, which enabled producers to distribute sudden cost increases over a longer period. Clients who consumed electricity at the time when the marginal cost of production was the highest paid roughly the same price as clients who consumed electricity when the marginal cost was at its lowest. With the advent of competition, this is no longer the case: competition has introduced true pricing. Tariffs are applied in real-time (producers must submit their supply bids for the following day), which tends to heighten price volatility.

Indeed, prices can be highly volatile on the spot markets, incurring price risk for the different players and sparking the need for a new function–risk management. As with physical sales, price risk can be hedged in two ways.

• Transactions can be made over-the-counter. This is notably the case when there is no organised market.

● Transactions can be made in organised financial markets, which offer derivatives (futures and options). In general, these organised financial markets are created as a natural complement to organised spot markets.

The first example is illustrated by the market that existed until recently in England and Wales. Owing to the high levels of price volatility on the Pool, the majority of market participants have contractual relationships in the form of Power Purchase Agreements (PPA). These can be forward contracts where the supplier agrees to deliver a certain volume of electricity to the consumer in the future, at a price agreed in advance. The contract may also be a Contract for Differences (CFD), which is a purely financial contract. A supplier and a consumer agree on a price for electricity to be supplied over a given period (the strike price). If the Pool price is above the strike price, the consumer pays the supplier the difference between the two prices, while if the strike price is below the Pool price, the supplier pays the difference to the consumer. Options are also available.

The second example is illustrated by the Nord Pool. In addition to the spot market (Elspot), the Nord Pool comprises a financial market (Eltermin) with a view to offering operators heading and risk management services. Eltermin contracts comprise three major categories; forward contracts, futures contracts and options. These concepts are explained in detail in Chapter 14.

13.9   References

1 DIMSON, E.: 'The discount rate for a power station', *Energy Economics*, July 1989, **11**, (3)
2 AUERBUCH, S. *et al.*: 'Capital Budgeting, technological innovation and the emerging competitive environment of the electric power industry', *Energy Policy*, February 1996, **24**, (2)
3 HERTZ, D. B.: 'Risk analysis in capital investment', *Harv. Bus. Rev.*, Sept/Oct 1979, pp. 169–181
4 'Incorporating Risk and Uncertainty in Power System Planning', The World Bank Energy and Industry Department, paper 17, June 1989
5 BREALEY, R. A., and MYERS, S. C.: 'Principles of Corporate Finance' (McGraw Hill, 2000, 6th edn.)
6 'Electricity generating costs for plants to be commissioned in 2000', UNIPEDE, Paris, January 1994
7 SPINEY, P., and WATKINGS, G.: 'Monte Carlo Simulation techniques and electric utility resource decision', *Energy Policy*, February 1996, **24**, (2)
8 REUTLINGER, S.: 'Techniques for project appraisal under uncertainty' (The Johns Hopkins University Press, for the World Bank, 1970)
9 POUIQUEN, L.: 'Risk Analysis in Project Appraisal' (The Johns Hopkins University Press, for the World Bank, 1975)
10 NYOUKI, E.: 'Deregulation, competition, price volatility & demand for risk management on the European electricity market', *World Power*, 2000

Chapter 14

Risk management – in electricity markets

14.1 Introduction

Risk is the hazard to which we are exposed because of uncertainty [1]. Risk is also associated with decisions. Where there are no uncertainties and no alternatives, there is no risk. Decisions that we make can affect these uncertainties and can reduce hazards.

Under regulated markets there were fewer uncertainties and risks because tariffs were almost fixed. With deregulation, electric power markets are volatile, prices change within a short time, and risks can be serious. In the deregulated market of electric power, electricity is a commodity and consumers have choices, correspondingly generators are competing among each other. This has created opportunities mainly for consumers and risks for producers, which need to be hedged.

Trading refers to transactions that take place directly between two parties or through an organised exchange. Although commodities trading in agricultural products has been established since the middle of the 19th century, it was only in 1996 that electricity future started to be traded in the New Year Mercantile Exchange in the US.

14.2 Qualifying and managing risk

We have to start by trying to quantify risk and methods for reducing and managing such risks by hedging or other means. This needs understanding of the risk management terminology and jargon, which is summarised below.

Robustness, the most fundamental measure of risk, is the likelihood that a particular decision will not be regrettable. If the decision-maker's choice turns out to be optimal no matter what nature chooses, his choice is robust. (Use of the term 'optimal' implies that the decision-maker has a single objective. Robustness is also defined in multiple-objective situations.) More often, his choice is optimal only for a subset of nature's possible outcomes. Suppose that there is a probability of 0.60 that an outcome or realisation materialises. Then the choice is robust with probability 0.60.

Exposure is a measure of loss if an adverse materialisation of uncertainties occurs for a particular choice. Sometimes exposure can be measured in dollars, but often not. It is difficult to attach a dollar value to loss and inconvenience for a power breakdown.

A hedge is an option that reduces risk. *Derivatives* are instruments to achieve this hedging. A derivative is a financial instrument (such as futures or options contract). It is so-called because it derives its value from a related or underlying asset. An underlying asset is the specific asset on which a financial instrument is based. Derivatives are not securities at all. They are just agreements between two parties with opposite views on the market to exchange risk. For the hedger, derivatives ultimately help protect against price increases and assist in lowering funding costs and obtaining better currency exchange rates in the international financial markets. Derivatives are a form of insurance against market swings that affect the value of underlying assets. Examples of derivative instruments are forwards, futures, options, and swaps [2,3].

Traded options – an option gives the firm the right (but not the obligation) to buy or sell an asset in the future at a price that is agreed upon today. There is a huge volume of trading in options that are created by specialised options exchanges. For example, you can deal in options to buy or sell common stocks, bonds, currencies, and electric power as a commodity.

Futures – a futures contract is an order that you place in advance to buy or sell an asset or commodity like electricity. The price is fixed when you place the order, but you do not pay for the asset until the delivery date. Futures markets have existed for a long time in commodities such as wheat, soyabeans, and copper. The major development in the 1990s occurred when the futures exchanges began to trade contracts on energy and electricity (see Section 14.2.2).

Forwards – futures contracts are standardised products bought and sold on organised exchanges. A forward contract is a tailor-made futures contract that is not traded on an organised exchange. The principal business in forward contracts occurs in the foreign exchange market, where firms that need to protect themselves against a change in the exchange rate buy or sell forward currency through a bank. Such forward contracts are increasingly being used in electrical power.

Traders in derivatives are usually referred to as hedgers. *Hedging* is taking a position in one security or asset to offset the risk associated with another security or asset. They are meant to gain protection and have some control over risk. Hedging instruments include forward sales contracts, future concepts, swaps and options. Hedging is meant to reduce risk and improve returns. This is achieved by balancing the gain/loss (in the physical market with the loss/gain in the derivative market).

Derivatives are the best way to hedge against adverse risk. Two types of hedging schemes allow protection in falling or raising markets.

Short hedge – a short hedge is a commitment to sell a product in the future at an agreed price. A decline in cash market prices is offset by profits in the futures market. The opposite applies when prices rise. A short hedge offers protection against falling prices.

Long hedge – a long hedge involves a commitment to purchase a commodity in the future at a fixed price. A long hedge locks in the price of the commodity (gas, electricity). It offers protection against rising prices.

14.2.1 Forward contracts

A forward contract is an agreement between two parties to buy or sell an asset at a certain future time for a specified price.

Price for a forward contract is

$$F = S\,e^{r(T-t)},$$

where F = forward price, S = price of the asset at time t, K = delivery or strike price, T = time of maturity of the contract, t = current time, r = risk-free rate of the economy.

Figure 14.1 depicts the profit/loss of a forward contract. The payoff depends on whether the investor took a long or short position. For a long position, the forward contract has a profit if the price S_T is larger than the strike price K. The contract ends with a loss otherwise. In contrast, for a short position, the contract makes a profit if the price S_T is smaller than the strike price K. This is because the writer of the contract receives a price higher at delivery than the current market price.

The forward contract allows customised expiration dates and requires no performance bond.

14.2.2 Futures contract

A futures contract is an agreement between two parties that has standardised terms and conditions, where the seller guarantees to deliver an asset at a specified price at a specified time in the future and the buyer guarantees the acquisition of that asset at the specified price.

Prices of future contracts are calculated with the same formula above. Therefore both future and forward priced contracts are similarly priced. The future price of the derivative depends on the underlying price of the asset, the maturity date of the contract and the risk-free rate of the economy. In ideal situations and under the same terms, futures prices are equal to forward prices. However, future contracts differ from forward contracts, because future contracts have specified maturity dates whereas a forward contract could have any maturity date. With a forward contract the outcome is usually currency cash while in futures contract it is a physical deliverable. Also, futures contracts require a down payment (like a performance bond) forward contracts require no performance bond.

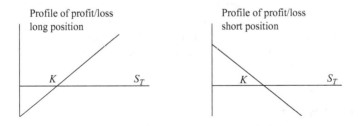

Figure 14.1   Profit of long and short positions

The following example is kept simple with the purpose of illustrating how a futures contract is used to hedge the risk in a rising market and how to avoid or minimise one's losses in a declining market.

An energy company sells electricity in the spot market. The company sells blocks of 500 MW at an hourly price.

(a) Assumption I: the party sells a futures contract (short hedge) to protect against falling markets.

(1) *Electricity prices fall*. Next month, the energy company will be selling 500 MWh in the spot market. The current price is $15 MWh^{-1} and in order to lock in this price, it sells a futures contract of 500 MWh through the exchange at $15 MWh^{-1} for the next month. The company targets revenue of $7500.00.

The electricity prices fall to $12.5 MWh^{-1}. The company receives from the spot market (500) ($12.5) = $6250.00. At the same time, the company buys back the futures contract that had a strike of $15 at the new market value of $12.5, realising a gain of ($2.5)(500) = $1250.00. Total revenue is $7500.00.

Following the market conditions, the company incurs a loss in the physical market, which is offset by a gain in the futures market.

(2) *Electricity prices rise*. The electricity prices rise to $17.5 MWh^{-1}. The company buys back the futures contract that had a strike set at $15 MWh^{-1} at a loss of $1250.00 and sells the 500 MWh on the commodity market for $8750.00. The total revenue is $7500.

(b) Assumption II: the party buys a futures contract to lock in a future price.

(1) *Electricity prices rise*. An energy broker commits to buy 500 MWh for February delivery from an energy producer. The broker enters into a futures contract for $15 MWh^{-1} for a total cost of $6250.00.

The energy company buys energy from the spot market at $17.5 MWh^{-1} and sells the electricity futures contract for $17.5 MWh^{-1}. The energy producer makes a $2.5 MWh^{-1} profit in the futures market, which offsets his loss in the physical market.

(2) *Electricity prices fall*. A Large Industrial User commits to buy 500 MWh of firm power that follows the spot market price for the next month. The user buys a futures contract for $15 MWh^{-1} for a total cost of $6250.00.

The electricity prices fall to $12.5 MWh^{-1}. The Large Industrial User buys electricity directly from the spot market. The Large Industrial User sells its futures contract at a loss of $2.5 MWh^{-1}.

The loss in the futures contract is offset by the gain in the spot market.

14.2.3 Options

An option is a type of contract that gives the purchasing party the right to buy or to sell for a certain price (called the exercise or strike price) and at a certain date (called the expiration date T, exercise date, or maturity date). The right to perform a financial transaction has a financial value called a premium. The buyer of the option pays a premium for the right (but not the obligation) to buy or sell the underlying asset (unlike the right or obligation to buy or sell a forward or futures contract, options are more versatile). Only a small percentage (approximately 30 per cent) of all purchased option contracts is exercised, the remaining bulk of which serves to hedge the exposure to the changes in the market price.

14.2.4 Swaps

A swap is an agreement between two parties to exchange one set of cash flows based on a notional amount linked to a floating index in exchange for another set of cash flows based on the same notional amount linked to a fixed index.

Swaps have revolutionised financial risk management and represent one of the greatest business opportunities in this area. Following the futures market in popularity, swaps are greatly utilised in the commodity markets; unlike the participants in a derivative exchange who have no knowledge of their counterparties and are willing to extend credit to them over a period of time.

There are fixed-fixed, fixed-floating and floating-floating swaps. The most common ones are fixed-floating swaps, which help a company that takes a loan at an adjustable interest rate to lock in the rate and thus eliminate any risk. The underlying asset of a swap could be a currency, interest rate, equity, or commodity.

Fixed to floating commodity swaps are common. An example will be a mining company that supplies coal to a generation utility. The mining company enters into an annual swap contract where it agrees to receive from the generation utility monthly payments based on delivery of 1000 tons of coal per month at the rate of $1 per one million BTU ($28 ton^{-1}).

In such an arrangement the mining company delivers coal each month with the spot market price per ton multiplied by 1000 and each month the company receives $28 multiplied by 1000.

Such an arrangement allows the mining company to receive fixed amounts. Simultaneously the mining company will deliver its coal at spot market value, which translates into negative or positive cash flow, as the spot market value is in excess or less than the fixed amount of $28 000.

There are several types of swap.

Commodity swap. This is an agreement between two parties to exchange cash flows based upon the difference between the fixed and floating prices of the same underlying principal. The exchange is done in a common currency.

Interest rate swap. One party agrees to pay to the other party a cash flow equal to a fixed interest rate on a notional amount. The exchange is done in the same currency. This is the most common swap.

Currency swap. This form of swap involves exchanging principal and fixed-rate interest payments on a loan in one currency for principal and fixed-rate interest payment on an approximately equivalent loan in another currency.

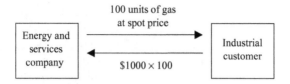

Figure 14.2   Fixed-floating commodity swap contract

Differential swap. This is a swap where a floating interest rate in the domestic currency is exchanged with a floating interest rate in a foreign currency as applied to the same notional principal.

There are endless combinations of swaps and other derivatives.

An example of a fixed-floating commodity swap contract is shown here. Suppose that an energy and services company supplies energy to a large industrial customer, in the form of 100 gas units each month for the next five years. The energy company enters into a semi-annual swap contract where it agrees to receive from the industrial customer a fixed monthly charge at the price of $1000.00 per unit in return for 100 gas units supplied. This swap is illustrated in Figure 14.2.

The exchange of payments takes place as follows:

- the company delivers gas each month with unit spot market value multiplied by 100 units,
- the company receives $1000.00 multiplied by 100 units each month.

This swap allows the company to receive a fixed amount, and in return the company will deliver its product at spot market value which translates into negative or positive cash flow as the current market value is in excess of, or less than, the fixed amount of $100,000.00.

Typically, this type of transaction is made through a financial institution that has the role of transforming a floating price (spot market) into a fixed value. Hedging allows the risk manager to reduce the risk exposure in highly volatile markets. Futures and options are often-used tools to manage financial risk.

14.2.5   Political risk

Political risk is very difficult to hedge. Such risk can come through changes in legislation and environmental agendas. In some cases, as in environmental legislation for SO_2 reduction where there is market for emission permits, hedges can be constructed. The same prospects exist for possible carbon taxation legislation. Political legislation for closing nuclear plants is a major political risk.

14.3   Decision making

A large part of engineering work involves making decisions [2] of various kinds. In fact, decision making ranks with innovation in its importance to the design process. The aim of a decision-making process is to come up with a decision that is optimal in some sense – i.e., the most cost-effective, giving the longest life, involving the lowest operating cost, leading to the most aesthetically pleasing product, and so on. Often, when such decisions are made, the decision maker is confronted with the problem of uncertainty. Intuition and past experience play a prominent role in the engineering decision-making process, but as important as they are, they are often insufficient vehicles to bring us to the best decision. In such cases, a rational approach using probabilistic tools permits proper modelling and analysis of uncertainty.

In any decision analysis, the set of decision variables should first be identified and defined. In a decision-making context, the engineer identifies courses of action or alternatives. Next, a prediction is made of the result of each course of action. For alternatives, we predict the initial cost, the maintenance costs, the life, the risk of failure, and other important variables. Each course of action results in a selection of certain maintenance policy or, in terms of the decision analysis, results in certain outcome.

Before a decision is made, we attempt to decide on the desirability of each outcome by attaching some value to it. Finally, we make our decision by selecting the outcome with the greatest value; i.e., the entire decision-making process can be viewed as an optimisation exercise.

14.3.1   The decision-making process

Any decision-analysis problem can be broken into four steps: (1) structuring the problem, (2) assessing the possible impacts of the alternatives, (3) determining the value structure, and (4) synthesising the information obtained in steps (1) to (3) to evaluate and compare the alternatives. There are usually several issues that have to be addressed in any particular problem. In any specific case, the resolution of these issues may be crucial to success. Therefore, all of these issues have to be addressed in the decision model. Figure 14.3 summarises the basic steps in the decision-analysis

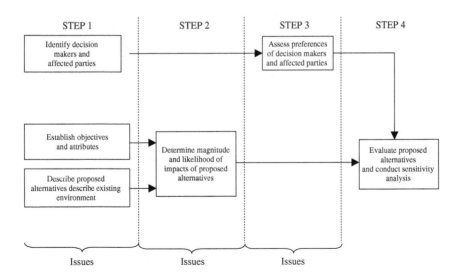

Figure 14.3   Schematic representation of the components of decision analysis [2]

approach and shows some major issues related to a technology choice in power systems [2,4].

14.4 References

1 PEREIVA, M., MCOY, M., and MERRILL, H.: 'Managing Risk in the New Power Business', *IEEE Computer Applications in Power,* April 2000, pp. 18–24
2 'Risk Assessment and Financial Management', *IEEE TP – 137 – Winter Meeting,* 1999
3 BREALY, R. A., and MYERS, S. C.: 'Principles of Corporate Finance', (McGraw-Hill, 2000, Sixth Edition)
4 KEENEY, R. L., BELEY, J. R., FLEISCHAUER, P., KIRKWOOD, C. W., and SICHERMAN, A.: 'Decision Framework for Technology Choice', Vol. 1: A case study of one utility's coal-nuclear choice, Electric Research Power Institute Report No. EA-2153, 1981

Appendix 1

Financial evaluation for choice of transformer

This is a real-life example of a detailed analysis for the least-cost choice of a transformer. It shows how the higher price of a facility can be traded against its operational cost over its life span. The example considers two offers for transformers. The first transformer offer is cheaper but it has higher losses, and slightly different payment conditions. The evaluation considers the overall cost of each alternative, its cash price plus its discounted cost of losses.

The two transformer offers have the following technical characteristics.

	Size (MVA)	Voltage (kV)	Losses (kW)	
			iron	copper
Transformer A	40	132/33	55	400
Transformer B	40	132/33	76	360

Their quoted prices and payment conditions are as follows.

	Price (£1000s)	Payments (£1000s)	
		on contract	on delivery
Transformer A	450	225	225
Transformer B	470	70	400

In both cases the delivery is one year after contract. Commissioning is six months after delivery. The transformers are assumed to be loaded at 50 per cent of full load at the first two years of service, and at 75 per cent of full load in the following two years. Afterwards it is fully loaded.

In this example, the expected life of the project is 30 years and a discount rate of 10 per cent is considered.

In such a case and in order to choose the least-cost solution it is required to consider the total cost of the project over its expected life span. This includes the price of the two transformers plus their discounted cost of the losses: fixed losses (iron losses) and load losses (copper losses). It is assumed that reliability of the transformers will be the same, also their maintenance costs; therefore these were ignored from the comparison.

It has to be realised that copper losses at any time are equal to full load copper loss kW (denoted C) multiplied by the square of the transformer's utilisation factor, i.e.

$$C = \left(\frac{\text{demand}}{\text{rated capacity}}\right)^2.$$

Therefore in the case of the first two transformers their combined annual copper losses at half load are:

$$400\,\text{kW} \left(\frac{20\,\text{MVA}}{40\,\text{MVA}}\right)^2 = 100\,\text{kw}.$$

To work out the annual energy losses, the load factor and shape of the load curve plays a significant role. Usually, and for most systems, average losses equal to peak losses multiplied by $(0.15 + 0.85 \times \text{load factor})$. With a load factor of 60 per cent this will be equivalent to 0.66.

Therefore annual losses at half load for the first two transformers are equal to

$$100\,\text{kW} \times 8760\,\text{h} \times 0.66 = 578\,160\,\text{kWh}, \text{ say } 578\,\text{MWh}.$$

In order to compute the value of the losses, these have to be multiplied by the marginal cost of energy at the substation site. This may be significantly lower than the tariff if the transformer is part of the utility system. For simplicity, a two-part energy cost is considered: $3.5\,\text{p kWh}^{-1}$ during the 16 h of peak and $2\,\text{p kWh}^{-1}$ for the off-peak 8 h. Also as an approximation, costs of all copper losses were calculated at the peak rate since the vast majority of such losses occur during these 16 h, and therefore no splitting of copper losses between peak and off-peak hours was undertaken.

Alternative A Transformers

Year	Cost (£1000s)	Losses (MWh)			Total losses (MWh)	
		Iron peak	Iron off-peak	Copper	Peak	Off-peak
−1	225					
0	225					
1		321	161	578	799	161
2		321	161	578	799	161
3		321	161	1300	1621	161
4		321	161	1300	1621	161
5		321	161	2313	2634	161
⋮						
29		321	161	2313	2634	161
30		321	161	2313	2634	161

For the sake of discounting, the base date (year) is considered to be the date of delivery of the transformers. For accuracy purposes, the losses have to be grouped at the middle of the year, and since the transformers will be operational six months from delivery, this means that losses can be grouped annually with respect to the base date (year), with the first year losses occurring one year after year 0.

It is much easier, to discount kWh and then multiply the discounted value of these by the cost. This is what has been done in the two following streams, each stream detailing the discounted total life span cost of each transformer alternative.

Alternative B Transformers

Year	Cost (£1000s)	Losses (MWh)			Total losses (MWh)	
		Iron peak	Iron off-peak	Copper	Peak	Off-peak
−1	70					
0	400					
1		444	222	520	964	222
2		444	222	520	964	222
3		444	222	1170	1614	222
4		444	222	1170	1614	222
5		444	222	2081	2525	222
⋮						
29		444	222	2081	2525	222
30		444	222	2081	2525	222

Transformer A

Annuity factor for 30 years at 10 per cent discount factor = 9.427
Annuity factor for 4 years at 10 per cent discount factor = 3.170
Annuity factor for 2 years at 10 per cent discount factor = 1.736
Annuity factor for the period 5–30 years at 10 per cent discount factor = 6.257

Discounted losses

Off-peak $\quad$ = 161 × 9.427 = 1518 MWh
Copper (peak) $-$ 578(1.736) | 1300(3.170 − 1.736)
$\qquad\qquad$ +2313 × 6.257
$\qquad\qquad$ = 1003 + 1864 + 14 472
$\qquad\qquad$ = 17 339 MWh
Total (peak) $\quad$ = 17 339 + fixed peak losses (321 × 9.427)
$\qquad\qquad$ = 20 367 MWh

Cost of losses

Cost of peak rate losses $\quad$ = 20 367 × 10^3 × 3.5 × 10^{-2} = £712 845
Cost of off-peak rate losses = 1518 × 10^3 × 2 × 10^{-2} = £30 360
Total cost of losses $\qquad$ = £743 205

Total cost of project $\qquad$ = 225 000 + (225 000 × 1.1) + 743 205
$\qquad\qquad$ = £1 215 705

Transformer B

Discounted losses

Off-peak $\quad$ = 222 × 9.427 = 2093 MWh
Copper (peak) = 520(1.736) + 1170(3.170 − 1.736)
$\qquad\qquad$ +2081 × 6.257
$\qquad\qquad$ = 903 + 1678 + 13021
$\qquad\qquad$ = 15 602 MWh
Total (peak) $\quad$ = 15 602 + (444 × 9.427) = 19 788 MWh

Cost of losses

Cost of peak rate bases $\quad$ = 19 788 × 10^3 × 3.5 × 10^{-2} = £692 580
Cost of off-peak rate losses = 2093 × 10^3 × 2 × 10^{-2} = £41 860
Total cost of losses $\qquad$ = £734 440

Total cost of project $\qquad$ = 400 000 + (70 000 × 1.1) + 734 426
$\qquad\qquad$ = £1 211 440

Conclusion

From the above analysis it is clear that the life span cost of the first alternative Transformer A is £1 215 705, while that of Transformer B is £1 211 426. Therefore Transformer B is the least-cost alternative, although it is slightly more expensive.

This real-life example shows how operational costs of almost all energy consuming facilities in the electrical power system play a more significant role in the overall life span cost than the investment cost. In this case the operational cost is almost twice that of the price and therefore more important in influencing the outcome. This consideration is still more important in case of generating facilities as shown in Chapter 5.

An alternative approach

A less accurate, but much more simplified, approach will be possible if the loading on each alternative transformer is assumed to be constant and at full load.

Utilising the Annual Cost Method (the Equivalent Uniform Annual Cost Method) of Chapter 5, the annual cost of each alternative will be as follows.

Transformer A

Capital cost × annuity factor + cost of losses
Copper losses	=	400 kW × 8760 h × 0.66
	=	2313 MWh
Iron losses	=	55 × 8760
	=	482 MWh (one third of these, i.e. 161 MWh, at the off-peak rate of 2 p kWh^{-1}.
Cost of losses	=	$(2313 + 321) \times 10^3 \times £0.035 + 161 \times 10^3$ ×£0.02
	=	92 190 + 3220 = £95 410
CRF	=	1/annuity factor = 1/9.427
	=	0.106
Annual cost of the investment	=	£450 × 10^3 × 0.106
	=	£47 700
Total annual cost	=	£143 110

Transformer B

Copper losses	=	360 kW × 8760 h × 0.66
	=	2081 MWh
Iron losses	=	76 kW × 8760
	=	666 MWh (one third of these, i.e. 444 MWh, at the off-peak rate of 2 p kWh^{-1})
Cost of losses	=	$(2081 + 444) \times 10^3 \times £0.035 + 222 \times 10^3$ ×£0.02
	=	£88 375 + £4440
	=	£92 815
Annual cost of the investment	=	£470 × 10^3 × 0.106
	=	£49 820
Total annual cost	=	£142 635

The same result is achieved, which again shows that Transformer B is the cheaper alternative. However, this method makes many approximations. The present value method is the more accurate and should be employed whenever possible.

Appendix 2

Glossary

The following is a glossary of financial, economic and technical terms used in this book as well as few other terms that are of interest. Sources for definition are in references [1], [2] and [6] of Chapter 3.

Amortisation   Gradual repayment or writing off of an original amount. Depreciation is a form of amortisation. The Capital recovery factor (see below) is composed of an interest component and an amortisation component.

Annual equivalent   A stream of equal amounts paid or received annually for a period such that by discounting at an appropriate interest rate it will have a specified present worth. Determined by multiplying an initial value by the capital recovery factor for the appropriate interest rate and period. To 'annualise' is to find the annual equivalent of a value.

Annuity   Investment that produces a level stream of cash flows for a limited number of periods.

Annuity factor   The present value at a discount rate r of an annuity of £1 paid at the end of each n periods:

$$\text{annuity factor} = \left[\frac{1}{r} - \frac{1}{r(1+r)^n} \right].$$

The annuity factor is obtained from annuity tables in the Financial References.

Appraisal   Analysis of a proposed investment to determine its merit and acceptability in accordance with established decision criteria (mostly carried out by banks and developmental agencies).

Arbitrage   Purchase of one security and simultaneous sale of another to give a risk-free profit.

Basis risk   Residual risk that results when the two sides of a hedge do not move exactly together.

Benefit–cost ratio (B/C ratio)   Discounted measure of project worth. The present worth of the benefit stream divided by the present worth of the cost stream. Often abbreviated to 'B/C', also frequently called the 'cost–benefit ratio'.

Benefits   The incremental value of product sales or cost reductions attributable to an investment. In case of the ESI, benefits can sometimes be represented in the form of kilowatt-hours (kWh).

Beta   Measure of market risk.

Bond   Long-term debt.

Border price   The unit price of a traded good at a country's border. For exports, the free-on-board (FOB) price; for imports, the cost–insurance–freight (CIF) price.

Break-even point   The level of product sales at which financial revenues equal total costs of production. At higher volumes of production and sales financial profits are generated.

Call option   Option to buy an asset at a specified *exercise price* on or before a specified exercise date (cf. *put option*).

Cap   An upper limit on the interest rate on a *floating-rate note.*

Capital budget   List of planned investment projects, usually prepared annually.

Capital rationing   Shortage of funds that forces a utility to choose between projects.

Capital recovery factor (CRF) or equivalent annual cost   Can be used to convert a sum of money into an equivalent series of equal annual payments, given a rate of interest and total period of time. The CRF for £1 at an interest rate of 12 per cent and a period of 4 years is £0.3292, as shown in the following schedule.

	Principal amortisation +	Interest =	CRF
	Principal	Interest	CEF (total)
1	0.2092	0.1200	0.3292
2	0.2343	0.0949	0.3292
3	0.2624	0.0668	0.3292
4	0.2941	0.0351	0.3292
	1.0000	0.3168	

The use of the CRF to compute annual capital charges is generally superior to the use of accounting depreciation allowances and interest expenses, for project appraisal purposes.

CAPM   Capital asset pricing model.

Cash flow   In its simplest concept, this is the difference between money received and money paid out. As used in benefit–cost studies, the net benefit stream anticipated for a project. Net benefits are available for the service of borrowed funds (amortisation, interest, and other charges), payments of dividends to shareholders, and the payment of profit taxes.

CIF   The landed cost of an import (cost, insurance, and freight) on the receiving country's dock, including the cost of international freight charges and insurance, before the addition of domestic tariffs or other taxes and fees.

Compound interest   Reinvestment of each interest payment of money invested, to earn more interest (compare with *simple interest*).

Compounding   The process of finding the future value in some future year of a present amount growing at compound interest.

Constant prices or **real prices**   Prices that have been adjusted to remove general price inflation.

Contingency allowance   An amount included in a project account to allow for adverse conditions that will add to baseline costs. Physical contingencies allow for physical events and unexpected costs, they are included in both the financial and the economic analysis. Price contingencies allow for general inflation; in project analysis they are omitted from both the financial and the economic analysis when the analysis is done in constant (real) prices.

Continuous compounding   Interest compounded continuously rather than at fixed intervals.

Covariance   Measure of the comovement between two variables.

Covenant   Clause in a loan agreement.

Conversion factor (standard conversion factor)   A number, usually less than one, that can be multiplied against a domestic market price of an item to reduce it to an equivalent border price. The simple version of the standard conversion factor is the ratio between a country's foreign trade turnover before and after import and export taxes (or subsidies).

Correlation coefficient   Measure of the closeness of the relationship between two variables.

Cost of capital   Opportunity cost of capital.

Costs   Costs are incurred to acquire project inputs such as buildings, electrical facilities and machines, materials, labour, and utilities. In economic evaluation certain outlays, such as the payment of profit taxes, are costs to the project but not the country. Such outlays are properly treated as transfers of project surplus rather than costs for the purpose of calculating net present value or internal rate of return.

Covariance   Measure of the comovement between two variables.

Crossover discount rate   The rate of discount that equalises the net present value of benefit or cost streams. Often applied to the cost streams of mutually exclusive project proposals. At a lower rate of discount, 'A' is superior, whereas at a higher rate of discount, 'B' is superior.

Current prices or **nominal prices**   Prices that have not been adjusted (deflated to eliminate general price inflation).

Cut-off rate   A rate of return established as a 'threshold' below which projects should not be accepted. See *Opportunity cost of capital.*

DCF   Discounted cash flow.

Debt service   A payment made by a borrower to a lender. May include one or all of: (1) payment of interest; (2) repayment of principal; and (3) loan commitment.

Decision tree   The diagram used in an analytical technique by which a decision is reached through a sequence of choices between alternatives. It is also a method of representing alternative sequential decisions and the possible outcomes from these decisions.

Deflation   The act of adjusting current to constant prices. The arithmetic (division) is the same as for discounting.

Depreciation   The anticipated reduction in an asset's value brought about through physical use or gradual obsolescence. Various methods are used: straight line, declining balance, accelerated, etc. The important thing to remember is that depreciation charges do not represent cash outlays and should not be included in financial or economic cash flows.

Discount factor   How much 1 at a future date is worth today. Also called the 'present worth factor' and the 'present worth of 1'.

Discount rate   A rate of interest used to adjust future values to present values. Discounting a future value to the present is the exact opposite of compounding a present value forward to a future value.

Discounted cash flow analysis   Analysis based on the net incremental costs and benefits that form the incremental cash flow. It yields a discounted measure of project worth such as the net present worth, internal rate of return, or net benefit–investment ratio.

Discounting   The process of finding the present worth of a future amount.

Distortion   A distortion exists when the market price of an item differs from the price it would bring in the absence of government restrictions.

EBIT   Earnings before interest and taxes.

Economic prices   Also known as 'efficiency' prices. Prices believed to reflect the relative scarcity values of inputs and outputs more accurately than market prices, owing to the influence of tariffs and other distortions in the latter.

Economic rate of return   The internal rate of return of a cash flow expressed in economic prices. Reduces the net present value of the cash flow to zero.

Equity   An ownership right or risk interest in an enterprise.

Exercise price *(striking price)*   Price at which a *call option* or *put option* may be exercised.

Expected return   Average of possible returns weighted by their probabilities.

Externality   In project analysis, an effect of a project felt outside the project and not included in the valuation of the project.

Factor of production   The inputs required to produce output. Primary factors of production are land, labour, and capital; secondary factors include materials and other inputs.

Financial leverage (gearing)   Use of debt to increase the *expected return* on *equity*. Financial leverage is measured by the ratio of debt to debt plus equity (operating leverage).

Financial prices   Synonymous with market prices.

Financial rate of return   The internal rate of return of a cash flow expressed in market prices. Reduces the net present value of the cash flow to zero.

First-year return   An analytical technique to determine the optimal time to begin a proposed project. The optimal time to begin the project is the earliest year for which the incremental net benefit stream for a project begun in that year has a first-year return exceeding the opportunity cost of capital.

Fixed costs   Costs that do not vary with changes in the volume of output.

FOB   The 'free-on-board' price of an export loaded in the ship or other conveyance that will carry it to foreign buyers.

Hedging   Buying one security and selling another in order to reduce risk. A perfect hedge produces a riskless portfolio.

Income statement   A financial report that summarises the revenues and expenses of an enterprise during an accounting period. It is thus a statement that shows the results of the operation of the enterprise during the period. Net income, or profit, is what is left over after expenses incurred in production of the goods and services delivered have been deducted from the revenues earned on the sale of these goods and services.

Incremental   Refers to the change in the production or consumption of inputs and outputs attributable to an investment project. Measuring project benefits and costs on a 'with/without' incremental basis rather than a 'before/after' basis is essential.

Intangible   In project analysis, this refers to a cost or benefit that, although having value, cannot realistically be assessed in actual or approximate money terms.

Interest during construction (IDC)   Interest charges occurred during project execution and normally capitalised up to the point in time when the plant starts commercial operation. However, neither interest during construction nor operation is included in the internal rate of return calculation.

Internal rate of return (IRR)   A discounted measure of project worth. The discount rate that just makes the net present worth of the incremental net benefit stream, or incremental cash flow, equal zero.

Liabilities, total liabilities   Total value of financial claims on a firm's assets. Equals (1) total assets or (2) total assets minus net worth.

LIBOR (London interbank offered rate)   This is the interest rate at which major international banks in London lend to each other. LIBID is London interbank bid rate; LIMEAN is mean of bid and offered rate.

Marginal productivity of capital   The economic productivity or yield of the last available investment money spent on the least attractive project.

Money terms   The monetary prices of goods and services. Distinguished from real terms, which refer to the physical, tangible characteristics of goods and services.

Mutually exclusive projects   Project alternatives that provide essentially the same output; if one is done the others are not needed or cannot be done.

Net benefit   In project analysis, the amount remaining after all outflows are subtracted from all inflows. May be negative particularly in the early years of the project. The net cash flow.

Net present value (NPV)   The sum of discounted future benefits and costs at a stated rate of discount. An absolute measure of project merit.

Net working capital   Current assets minus current liabilities.

Nominal   Stated as an amount of money. Compare with *real*.

Non-traded   A project input or output that is not traded by a country either because of its production cost, bulkiness or because of restrictive trade practices.

Opportunity cost   Value lost by using something in one application rather than another. The opportunity cost of employing a worker in a project is the loss of net output that worker would have produced elsewhere. The concept of opportunity cost is the corner stone of benefit–cost analysis.

Opportunity cost of capital   The return on investments foregone elsewhere by committing capital on the project under consideration. Also referred to as the marginal productivity of capital, a rate of return that would have been obtained by the last acceptable project. The opportunity cost of capital is normally used as a 'cut-off rate' in investment decisions.

Option   See *call option, put option*.

Payback period   Time taken for a project to recover its initial investment in monetary terms.

P/E ratio   Share price divided by earnings per share.

Perpetuity   Investment offering a level stream of cash flows in perpetuity.

Present value   Discounted value of future cash flows.

Present worth (PW)   (1) The value at present of an amount to be received or paid at some time in the future. Determined by multiplying the future value by the discount factor. (2) The sum of the present worths of a series of future values.

Price elasticity   Price elasticity refers to the relationship between the percentage change in the quantity demanded or supplied of an item with respect to a stated percentage change in the item's unit price.

Cross elasticity refers to the influence of the price of one item on the demand for another. If a reduction in the price of X leads to increased demand for X and for Y declines, the two are 'competitive'. Zero cross elasticities indicate perfect complementarity, and infinite cross elasticities indicate perfect substitutability.

Pro forma   Projected.

Profit   Financial profit is the difference between financial revenues and costs. Economic profit is the surplus of benefits over costs when economic prices are used, after deducting the opportunity cost of capital.

Protection   Measures that protect domestic producers from foreign competitors, including import tariffs, quotas, and administrative restrictions that effectively limit or prevent foreign competition. Most accurately measured as the difference between border prices and market prices, after allowing for domestic transfer costs.

Put option   Option to sell an asset at a specified *exercise price* on or before a specified exercise date (cf. *call option*).

Rate of return   Remuneration to investment stated as a proportion or percentage. Often the internal rate of return. The financial rate of return is the internal rate of return based on market prices; the economic rate of return is the internal rate of return based on economic values.

Real interest rate   Interest rate expressed in terms of real goods, i.e. nominal interest rate adjusted for inflation.

Real prices   See *constant prices*.

Return on equity   (1) The internal rate of return of the incremental net benefit after financing. Used as a measure of project worth. (2) Net income divided by equity.

Return to capital   The rate of return received by the investor on capital engaged in a project.

Risk premium   Expected additional return for making a risky investment rather than a safe one.

ROI   Return on investment.

Salvage value (residual value)   Scrap value of plant and equipment.

Sensitivity analysis   A systematic review of the impact that changes in selected benefits and costs have on a project's net present value of internal rate of return.

Shadow price (accounting price)   A price that is computed rather than observed in a market place.

Simple interest   Interest calculated only on the initial investment.

Standard deviation   A measure of the dispersion of a frequency distribution. Obtained by extracting the square root of the arithmetic mean of the squares of the deviation of each of the class frequencies from the arithmetic mean of the frequency distribution.

Striking price   *Exercise price* of an *option*.

Sunk cost   A cost incurred in the past that cannot be retrieved as a residual value from an earlier investment.

Swap   An arrangement whereby two companies lend to each other on different terms, e.g. in different currencies, or one at a fixed rate and the other at a floating rate.

Traded   A project input or output is said to be traded if its production or consumption will affect a country's level of imports or exports, at the margin. A partially traded item will also affect the level of domestic production or consumption.

Transfer payment   A payment made without receiving any good or service in return.

Treasury bill   Short-term discount debt maturing in less than 1 year, issued regularly by the government.

Unique risk   (residual risk, specific risk, unsystematic risk) Risk that can be eliminated by diversification.

Variable costs   Costs that vary with changes in the level of output, such as costs for fuel.

Variance   Mean squared deviation from the expected value – a measure of variability. Standard deviation is equal to the square root of the variance.

Weighted-average cost of capital   Expected return on a portfolio of all the firm's securities. Used as hurdle rate for capital investment.

With and without   Refer to the situations with and without a proposed project. In project analysis, the relevant comparison is the net benefit with the project compared with the net benefit without the project. This is distinguished from a 'before and after' comparison because even without the project the net benefit in the project area may change.

Working capital   Current assets and current liabilities. The term is commonly used as being synonymous with net working capital.

Index

Printed in the USA
CPSIA information can be obtained
at www.ICGtesting.com
JSHW011509221024
72173JS00005B/1259

9 780863 413049